D1753712

Battaglia · Pfannhauser · Murkovic
Who's Who in Food Chemistry – Europe, 2nd Edition

Springer
*Berlin
Heidelberg
New York
Barcelona
Hong Kong
London
Milan
Paris
Singapore
Tokyo*

Reto Battaglia · Werner Pfannhauser
Michael Murkovic (Eds.)

Who´s Who in Food Chemistry

Europe

2nd completely rev. and exp. edition

Federation of European Chemical Societies
Division of Food Chemistry

Springer

Dr. Reto Battaglia
Central Laboratory
Federation of Migros Cooperatives
Hönggerstraße 24
CH-8031 Zurich

Prof. Dr. Werner Pfannhauser
Dr. Michael Murkovic
Department of Food Chemistry
and Technology
Graz University of Technology
Peterssgasse 12/2
A-8010 Graz

ISBN 3-540-41448-7 Springer-Verlag Berlin Heidelberg New York

CIP data aaplied for

Die Deutsche Bibliothek - CIP-Einheitsaufnahme
Who`s who in food chemistry : Europe / Reto Battaglia ... (ed.). - 2., completely rev. and expanded ed.. -Berlin ; Heidelberg ; New York ; Barcelona ; Hong Kong ; London ; Milan ; Paris ; Singapore ; Tokyo : Springer, 2001
ISBN 3-540-41448-7

This work is subject to copyright. All rights are reserved, whether the whole or part of the material is concerned, specifically the rights of translation, reprinting, reuse of illustrations, recitation, broadcasting, reproduction on microfilm or in other ways, and storage in data banks. Duplication of this publication or parts thereof is permitted only under the provisions of the German Copyright Law of September 9, 1965, in its current version, and permission for use must always be obtained from Springer-Verlag. Violations are liable for prosecution act under German Copyright Law.

Springer-Verlag Berlin Heidelberg New York
a member of BertelsmannSpringer Science+Business Media GmbH

http://www.springer.de

© Springer-Verlag Berlin Heidelberg 2001
Printed in Germany

The use of general descriptive names, registered names, trademarks, etc. in this publication does not imply, even in the absence of a specific statement, that such names are exempt from the relevant protective laws and regulations and therefore free for general use.

Typesetting: Dataconversion by author
Cover-design: Künkel + Lopka, Heidelberg
Printed on acid-free paper   SPIN: 10754685   52 / 3020 hu - 5 4 3 2 1 0 -

Preface

The FECS Division of Food Chemistry started the project "Who's Who in Food Chemistry – Europe" in 1996. After publication of the first edition, an increased interest of the European food chemists the delegates of the Food Chemistry Division decided to initiate an update of this compilation. The problem we faced was that a lot of the information presented in the first edition was no longer valid. Rapid changes of interests due to emerging new technologies in science and production and of course changes of mainly telephone numbers during the few years that had passed made a new edition of this book necessary.

The description of the specialization as well as the fields of interest were changed from a personal description to a long list of keywords. This should enable the reader to identify scientists working in a special area easily. Due to the increased use of electronic communication the email addresses were added.

This book will hopefully help to intensify the co-operation of the European food chemists and optimize the basic and applied research in this field.

GRAZ, FEBRUARY 2001   MICHAEL MURKOVIC

Contents

Countries 1
Personalities 7
Names 121
Keywords 129

Countries

Austria

Bauer, Friedrich 7
Böhm, Josef 7
Jirovetz, Leopold 7
Kadi, Andreas 7
Kroyer, Gerhard 7
Krska, Rudolf 7
Leibetseder, Josef 8

Leitner, Erich 8
Luf, Wolfgang 8
Murkovic, Michael 8
Pfannhauser, Werner 8
Schmid, Erich R. 8
Siegmund, Barbara 9
Sontag, Gerhard 9

Steiner, Ingrid 9
Tiefenbacher, Karl 9
Ulberth, Franz 9
Van Eckert, Renate 9
Wiedner, Peter 9
Winkler, Johanna 10
Weisz, Richard 10

Belgium

Balduck, Paul 10
Beernaert, Hedwig 10
De Block, Jan 10
De Brabander, Hubert 11
De Ruyck, Hendrik 11
Deelstra, Hendrik 11
Deweghe, Liane 11

Dirinck, Patrick 11
Herman, Lieve 11
Ilsbroux, Ingrid 12
Knowles, Michael Ernest 12
Ooghe, Wilfried 12
Schreyen, Luc 12
Steegmans, Monique 12

Temmerman, Guy 12
Van Peteghem, Carlos 12
Van Renterghem, Roland 13
Van Vyncht, Gery 13
Vanluchene, Eric 13

Bulgaria

Dalev, Pencho 13
Doncheva, Ivanka 13
Ivanov, Kalintcho 14
Kovatcheva-Apostolova, Elena 14

Neicheva, Anastasia 14
Obretonov, Tzvetan 14
Popov, Dimitre 14
Ribarova, Fanny 14
Rizov, Nicolay 14

Stanoeva, Elena 15
Topalova, Ivanka Chrstova 15
Tsanev, Roumen 15
Yanishlieva-Maslarova, Nedjalka 15

Croatia

Bazulic, Davorin 15
Galic, Kata 16
Gojmerac, Tihomira 16
Hardi, Jovica 16
Hasenay, Damir 16
Hegedusic, Vesna 16
Horvatic, Marija 16
Hruskar, Mirjana 17

Klapec, Tomislav 17
Kordic, Jasna 17
Kovac, Spomenka 17
Mandic, Milena L. 17
Novakovic, Predrag 17
Pilizota, Vlasta 18
Pospisil, Jasna 18
Pozderovic, Andrija 18

Primorac, Ljiljana 18
Sapunar-Postruznik, Jasenka 18
Sebecic, Blazenka 18
Subaric, Drago 19
Ugarcic-Hardi, Zaneta 19
Vahcic, Nada 19
Vedrina-Dragojevic, Irena 19

Czeck Republic

Benda, Vladimir 19
Bubnik, Zdenek 19
Cejpek, Karel 20
Chumchalová, Jana 20
Copíková, Jana 20
Culik, Jirí 20
Curda, Ladislav 20
Davidek, Jiri 20
Demnerova, Katerina 20
Dolezal, Marek 21
Dostálová, Jana 21
Drdak, Milan 21
Filip, Vladimir 21
Holandová, Katerina 21

Holasova, Marie 21
Hrncirik, Karel 21
Ingr, Ivo 22
Kalac, Pavel 22
Kas, Jan 22
Kellner, Vladimir 22
Kucera, Jiri 22
Kvasnicka, Frantisek 22
Masková, Eva 23
Melzoch, Karel 23
Mikova, Kamila 23
Panovská, Zdenka 23
Pavelka, Jiri 23
Pipek, Petr 23

Plocková, Milada 24
Pokorný, Jan 24
Poustka, Jan 24
Prugar, Jaroslav 24
Pudil, Frantisek 24
Rauch, Pavel 25
Rihakova, Zdenka 25
Rychtera, Mojmír 25
Schwarz, Walter 25
Stetina, Jiri 25
Suiraková, Eva 25
Valentová, Helena 25
Velisek, Jan 26
Vlater, Vladimír 26

Vodrazka, Zdenek 26

Denmark

Balling-Engelsen, Soren 26
Bjergegaard, Charlotte 26
Christophersen, Carsten 26
Holch, Klaus 27
Holmer, Gunhild 27

Melchior-Larsen, Lone 27
Munck, Lars 27
Ogaard-Madsen, Jorgen 27
Otte, Jeanette 27
Ross-Petersen, Karl Jakob 28

Skibsted, Leif H. 28
Sorensen, Susanne 28
Sorensen, Hilmer 28

Estland

Ilmoja, Kalle 28
Vokk, Raivo 28

Finland

Ahola, Maarit 29
Ahvenainen, Juha M.I. 29
Anderssen, Valborg 29
Applebye, Ulla 29
Aro, Tarja 29
Aura, Anna-Marija 29
Eerikäinen, Tero 29
Hägg, Margareta 30
Hakala, Mari 30
Häkkinen, Sari 30
Heiniö, Raija Liisa 30
Heinonen, Marina 30
Hietaniemi, Veli 30
Home, Silja D. 31
Honkavaara, Markku 31
Hopia, Anu 31
Horne-Ekman, Maarit 31
Huopalahti, Rainer 31
Järvenpää, Eila 31

Järvi-Kääriäinen, Irma
Terhen 31
Kallio, Heikki 31
Keurulainen, Ritva 32
Kivistö, Laura 32
Korhonen, Hann 32
Koskenkorva, Anneli 32
Kuusisto, Päivi 32
Laakso, Päivi 32
Lampi, Anna-Maija 32
Lampolahti, Soili 33
Lapveteläinen, Anja 33
Lehtonen, Pekka 33
Määttä, Kaisu 33
Mäkinen, Marjukka 33
Matilainen, Katri 33
Moilanen, Raija 33
Niemi, Sanna-Maria 33
Ollilainen, Velimatti 34

Piironen, Vieno 34
Rantamäkki, Pirjio 34
Räsänen, Janne 34
Richter, Timo 34
Saarinen, Niina 34
Salminen, Seppo 34
Saukko, Maire 34
Silanpää, Rauno 35
Skrökki, Leila Anneli 35
Storgards, Erna 35
Suutarinen, Marjaana 35
Tahvonen, Raija 35
Tiusanen, Sirkka 35
Toikkanen, Kari 35
Toivo, Jari 35
Tuomala-Saramäki, Terhi 36
Tykkyläinen, Paavo 36
Vahteristo, Liisa 36

France

Birlouez, Inès 36
Cheftel, Jean-Claude 36
Decaris, Bernard 36
Ducauze, Christian J. 37
Feinberg, Max 37
Gaucheron, Frédéric 37

Graille, Jean 37
Ilari, Jean-Luc 37
Le Botlan, Denis 37
Liddle, Peter 37
Marin, Michèle 38
Martin, Gérard 38

Meunier, Jean-Claude 38
O'Brien, John 38
Rahali, Véronique 38
Rollin, Patrick 38
Rutledge, Douglas 39
Talou, Thierry 39

Germany

Alder, Lutz 39
Aust, Olivier 39
Baltes, Werner 39
Bauer, Ulrich 39
Berger, Ralf Günter 40
Betsche, Thomas 40

Böhm, Volker 40
Brockmann, Rainer 40
Brockmann, Anneliese 40
Brunn, Hubertus 40
Budde, Jürgen 40
Büning-Pfaue, Hans 41

Christoph, Norbert 41
Daphi-Weber, Juliane 41
Dawihl, Gerd 41
Dechow, Arndt 41
Dettweiler, Gerd 41
Dietrich, Helmut 41

Eberle, Mike 42
Ehlers, Dorothea 42
Eisenbrand, Gerhard 42
Enders, Peter W. 42
Erning, Dieter 42
Franzke, Claus 43
Gertz, Christian 43
Glück, Bernfried 43
Haffke, Helma 43
Hahn, Harald 43
Hanewinkel-Meshkini,
 Susanne 43
Heinzler, Matthias 43
Henle, Thomas 44
Hey, Hanke 44
Hils, Arno K. A. 44
Honikel, Karl Otto 44
Janson-Mundel, Ortrun 44
Jörissen, Urban 44
Klein, Erich 45
Klostermeyer, Henning 45
Knieling, Ralph G. 45
Kombal, Ralph 45
Kramer, Jörg 45
Krause, Wolfgang 45
Kroh, Lothar W. 45
Kroll, Jürgen 46
Lach, Günter 46
Lambert, Michael 46

Landsiedel, Robert 46
Lautenbacher,
Lutz-Michael 46
Lindhauer, Meinolf G. 46
Littmann-Nienstedt, Sigrid 47
Lorenzen, Kay 47
Luckas, Bernd 47
Ludwig, Eberhard 47
Maier, Hans Gerhard 47
Malisch, Rainer 47
Malwitz, Dietmar 47
Marx, Friedhelm 48
Matissek, Reinhard 48
Mischnick, Petra 48
Mörsel, Jörg-Thomas 48
Mosandl, Armin 48
Nehring, Ulrich P. 48
Nöhle, Ulrich 49
Ochs, Stefan 49
Otteneder, Herbert 49
Petz, Michael 49
Pfalzgraf, Andreas 49
Pischetsrieder, Monika 49
Plaga-Lodde, Annette 49
Pollmer, Udo 49
Popken, Anne M. 50
Preuss, Axel 50
Ragotzky, Klaus 50
Rimkus, Gerhard G. 50

Ristow, Reinhard 50
Rohn, Sascha 50
Rüdt, Ulrich 51
Schäfer, Karola 51
Schlett, Claus 51
Schmidt, Heinz 51
Schmolck, Werner 51
Schneider, Rüdiger 51
Schreier, Peter 51
Schrenk, Dieter 52
Schulte, Erhard 52
Schwack, Wolfgang 52
Schwenke, Klaus Dieter 52
Seibel, Wilfried 52
Siebers, Johannes 52
Speer, Karl 53
Steinhart, Hans 53
Stelz, Alice 53
Tauscher, Bernhard 53
Ternes, Waldemar 53
Thier, Hans-Peter 53
Vieths, Stefan 54
Von Rymon Lipinski,
 Gert-Wolfhard 54
Weder, Jürgen, K.P. 54
Winkeler, Heinz-Dieter 54
Winterhalter, Peter 54
Wittenschläger, Lutz 54
Wucherpfennig, Karl 54

Greece

Adamantiadou, Sophia 55
Assimakopoulou,
 Angelique 55
Boskou, Dimitrius 55
Chroneos, Ioannis 55

Chrysafidis, Dimitrios 55
Demopolous,
 Constantinos A. 55
Economides, Anna 55
Komaitis, Michael 56

Kontominas, Michael 56
Marioleas, Panagiotis 56
Sakellariou, Christina 56
Scordaki, Alexandra 56
Tzia, Constantina 56

Hungary

Farkas, Joszef 57
Györi, Zoltán 57
Horváth-Mosonyi, Magda 57
Kovacs, Elisabeth Teresia 57
Lasztity, Radomir 57
Lugasi, Andrea 57

Molnar, Pal 58
Molnar-Perl, Ibolya 58
Salgò, Andràs 58
Sass-Kiss, Agnes 58
Simon-Sarkadi, Livia 58
Szilágyi, Szilárd 58

Tömösközi, Sándor 59
Tóth-Markus, Marianna 59
Varadi, Maria 59
Viszkok, Ferenc 59

Ireland

Byrne, Briege Eileen 59
Hood, Ted D.E. 59

Mc Donald, Mark 60
Roos, Yrjö Henrik 60

Spillane, William J. 60
Troy, Declan J. 60

Italy

Accorsi, Carla Alberta 60
Anklam, Elke 60
Berardo, Nicola 60
Bontempelli, Gino 61
Botrè, Claudio 61
Botrè, Francesco 61
Burini, Giovanni 61
Calzolari, Claudio 61
Campanella, Luigi 61
Chiavaro, Emma 62
Corradini, Claudio 62
Damiani, Pietro 62
Di Luccia, Aldo 62
Di Natale, Corrado 62

Dossena, Arnaldo 62
Evangelisti, Filippo 63
Evidente, Antonio 63
Favretto, Luciano 63
Ferrara, Lydia 63
Gabrielli Favretto, Luciana 63
Gatti, Gian Carlo 64
Iori, Renato 64
Luneia, Roberto 64
Marchelli, Rosangela 64
Marini, Domenico 64
Martelli, Aldo 64
Mazzei, Franco 65
Merlini, Lucio 65

Moret, Ivo 65
Palla, Gerardo 65
Pertoldi Marletta, Giuliana 65
Pisciotta, Gennaro 65
Restani, Patrizia 66
Salvo, Francesco 66
Sforza, Stefano 66
Tateo, Fernando 66
Uccella, Nicola Antonio 66
Valentini, Giuseppa 66
Ziino, Marisa 66
Zunin, Paola 67

Liechtenstein

Matt, Helmuth 67

Lithauania

Venskutonis, Petras Rimantas 67

Netherlands

Aalbersberg, Willem Y. 67
Baars, Aalbert Jan 67
Beentjes, Pieter 68
Beljaars, Paul 68
Bontenbal, Edwin 68
Clark, David 68
Cornelese, Johan 68
De Jong, Jacob 68
De Kruif, Kees C.G. 68
De Ruiter, Gerhard A. 69
Ellen, Geert 69
Eshuis, Dolf F. 69
Groothuis, Dirk G. 69
Havenaar, Robert 69
Hermus, Rudolph J.J. 69
Kan, Kees 69

Kerkvliet, Jacob 70
Klaarenbeek, Tineke 70
Langendam, Johannes 70
Lauwaars, Margreet 70
Leloux, Mirjam 70
Lossonczy Von Losoncz,
Thomas 70
Maclean, Wim 71
Northolt, Martin 71
Noteborn, Hubert 71
Olieman, Cornelis 71
Roel, Peter 71
Rombouts, Frank 71
Roozen, Jacques 72
Rops, Wichard 72
Ruiter, Adriaan 72

Schutte, Leonard 72
Siezen, Roland J. 72
Stark, Jacques 72
Steinbuch, Erwin 72
Timmermans, Eric 73
Uijttenboogaart, Theo 73
Van Boekel, Tiny 73
Van Den Bosch, Gerrit 73
Van Den Broek, Ad 73
Van Der Schee, Henk A. 73
Van Dokkum, Wim 74
Van Poppel, Geert A.F.C. 74
Van Rhyn, Hans 74
Voragen, Fons 74
Wijngaards, Gerrit 74
Zagt, Robert 74

Norway

Brathen, Gudmund 75
Eklund, Trygve 75

Nortvedt, Ragnar 75
Sarpeid, Hans-Jacob 75

Slinde, Erik 75
Sorhaug, Terje 75

Poland

Amarowicz, Ryszard 75

Bachman, Stefania 76

Bykowski, Piotr Jan 76

Sweden

Jägerstad, Margaretha 93
Lingnert, Hans 93
Österdahl, Bengt-Göran 93
Sandberg, Ann-Sofie 94

Switzerland

Battaglia, Reto 94
Bosset, Jacques Olivier 94
Corvi, Claude Albert 94
Galluser, Anita 94
Gaudard - De Weck, Daniele 94
Hauser, Eugen J. B. 95
Kaufmann, Anton 95
Kieffer, Felix 95
Klein, Bernard 95
Koch, Herbert 95
Löliger, J. 95
Mikuschka, Gerhard 96
Mlotkiewicz, Jerzy 96
Moser, Ulrich 96
Schlatter, Christian 96
Schmitt, Rudolf 96
Studer, Alfred 96
Van Dael, Peter 97
Vischer, Michaela 97
Von Wietersheim, Eugen 97
Zoller, Otmar 97

Turkey

Basman, Arzu 97
Celik, Süeda 97
Köksel, Hamit 97
Yalcin, Erkan 98

United Kingdom

Adams, J. Brian 98
Ager, Elaine 98
Ames, Jennifer 98
Anderson, Raymond 98
Andrews, Anthony 98
Ashurst, Philip Roy Baigrie, Brian 98
Baines, David Allan 99
Beddows, Clifford G. 99
Belton, Peter 99
Bhat, Mahalingeshwara 99
Birch, Gordon G. 100
Boast, Martin 100
Branch, Simon 100
Brechany, Elizabeth 100
Brown, Peter Anthony 100
Buglass, Alan J. 100
Bunn, Cheryl 100
Campbell, Duncan J. 100
Carder, John 101
Clark, Michael 101
Clutton, David 101
Cooper, Julian 101
Cotte, Virginie 101
Coveney, Leslie 101
Crossy, Paul 102
Davies, Robert John 102
Davies, Alan Philipp 102
Dickinson, Eric 102
Donald, Athene 102
Ennion, Ronald 102
Farmer, Linda 102
Faulds, Craig 102
Fenwick, Gruffydd Roger 103
Fillery-Travis, Annette 103
Finnegan, Derek 103
Finney, Graham 103
Fisher, Leonard 103
Flowerden, Mary 103
Frazier, Peter 103
Fryer, John 104
Gates, Leonard Michael 104
Gilbert, John 104
Goodall, David M. 104
Goodman, Bernard 104
Gordon, Michael 104
Gramshaw, J.W. 104
Grenby, Trevor Hilary 105
Gunstone, Frank 105
Hamilton, Colin A. 105
Hampton, Ian 105
Harris, Caroline A. 105
Henderson, Nick C. 105
Henshall, David 105
Hey, Michael James 106
Hills, Brian 106
Hitchcock, Christopher 106
Hodgson, Ian 106
Holst, Birgit 106
Horne, David S. 106
Hough, Leslie 106
Howard, Julie 107
Howick, Chris 107
Ibe, Frank 107
Jones, Arthur David 107
Jones, James 107
Jones, Alan 107
Khokhar, Santosh 107
Klupsch, Robert N. 108
Kroon, Paul 108
Lalljie, Sam 108
Lees, Ronald 108
Leigh, Anthony 108
Lohman, Joost A. B. 108
Long, Alan 108
Mac Dowall, James 109
Makris, Dimitris 109
Malbon, Raymond 109
Martin, Peter Gerard 109
Mills, Elizabeth Naomi Clare 109
Morgan, Michael R.A. 109
Morris, Victor John 109
Mottram, Donald 110
Murray, Brent Stuart 110
Nursten, Harry Erwin 110
O' Neill, Ian 110
Parker, Jane 110
Parr, Adrian James 111
Payne, Nigel Kenneth 111
Pearce, Steven 111
Pearce, J. 111
Ridgway, Christopher 111
Ridley, Brian 111
Robins, Elizabeth Naomi Clare 111

Roedig-Penman, Andrea 111
Russell, Wendy Roslyn 112
Saltmarsh, Mike 112
Shackleton, Ronald 112
Sheppard, Peter D. 112
Sherlock, John C. 112
Sime, John 112
Smith, Linda Bernhardine Margaret 112
Spiro, Michael 113
Spooner, Martin John Richard 113

Stevens, Roger 113
Talbot, Geoff 113
Tennant, David 113
Thompson, Kenneth Clive 113
Thomas, Mark A.L. 113
Thornton, Raymond E. 113
Varey, Jane Elizabeth 114
Vella, Anthony 114
Waldron, Keith 114
Wang, Rui 114
Webb, Colin 114
Whitehead, John A. 114

Whitehouse, Brian 115
Wiggins, Edgar Hugh 115
Wilde, Peter 115
Williams, Mervyn 115
Williams, John Graham 115
Williams, Peter 115
Williamson, Gary 116
Wilson, Peter D.G. 116
Wilson, Philip 116
Wilson, Reginald 116
Wood, Brian J.B. 116
Zabetakis, Ioannis 116

Yugoslavia

Estelecki, Ilona 116

Personalities

Austria

BAUER, Friedrich Dipl.-Ing., Dr.techn.
University of Veterinary Medicine
Department of Meat Hygiene, Meat Technology and Food Science
Associate Professor
Veterinärplatz 1
A - 1210 Vienna
Tel :+43 1 250773302
Fax: +43 1 250773390
E-Mail: friedrich.bauer@vu-wien.ac.at

Private address
Amtsstraße 13
A - 3140 St. Pölten
Tel: +43 2742 42240
Fax: +43 2742 42240

Fields of interest
Meat and meat products, poultry, additives, proteins, biogenic amines, electrophoresis, HPLC, sensory analysis, adulteration, analytical chemistry, quality, storage

BÖHM, Josef Dr.
University of Veterinary Medicine
Institute of Nutrition
Veterinärplatz 1
A - 1210 Vienna
Tel :+43 1 25077 3215
Fax: +43 1 25077 3290
E-Mail: josef.boehm@vu-wien.ac.at

Private address
Schimmelgasse 14/9
A - 1030 Vienna
Tel: +43 1 7136034

Fields of interest
Meat and meat products, poultry, oils and fats, dietary fibres, lipids and fatty acids, mycotoxins, plant toxins, GC, HPLC, MS, bioavailability, metabolism, nutrition, biotechnology, fermentation, hygiene

JIROVETZ, Leopold Mag.pharm., Dr.rer.nat.
University Vienna, Dept. of Pharm. Chem.
Head of Laboratory
Althanstraße 14
A - 1090 Vienna
Tel :+43 1 313368641, 313368333
Fax: +43 1 31336771
E-Mail: jirovetz@speedy.pch.univie.ac.at

Private address
Ungerndorf 38
A - 2133 Ungerndorf/Loosdorf
Tel: +43 2522 88126

Fields of interest
Cosmetics, fruits and vegetables, spices, wine, aroma active compounds, lipids and fatty acids, Maillard reaction products, steroids, drugs, heterocyclic aromatic amines, chemometrics, GC, HPLC, MS, sensory analysis, spectroscopy, TLC, adulteration, analytical chemistry, food composition, nutrition, toxicology, fermentation, preservation, storage

KADI, Andreas M.Sc., MBA
Coca-Cola GmbH
Director Scientific & Regulatory Affairs Central and Eastern Europe
Triester Str. 217
A - 1230 Vienna
Tel :+43 1 66717159
E-Mail: ankadi@eur.ko.com

Private address
Glockengasse 29
A - 1020 Vienna

Fields of interest
Beverages, additives, consumer research, management, nutrition, quality, regulative issues, quality assurance, HACCP

KROYER, Gerhard Th. Univ. Prof., Dipl.-Ing., Dr.techn.
Technical University Vienna
Institute of Food Chemistry and Technology
Getreidemarkt 9
A - 1060 Vienna
Tel :+43 1 5880116003
Fax: +43 1 5880116099

Private address
Franz-Liszt-Gasse 28
A - 7092 Winden/See

Fields of interest
Cosmetics, sweeteners, additives, antioxidants, colours, polyphenols, analytical chemistry, bioavailability, nutrition, quality

KRSKA, Rudolf Prof., Dr.
Centre for Analytical Chemistry
Head of Department
IFA-Tulln, Konrad Lorenzstr. 20
A - 3430 Tulln
Tel :+43 2272 66280401
Fax: +43 2272 66280403
E-Mail: krska@ifa-tulln.ac.at

Private address
Dr. Lorenz-Böhm-Gasse 6
A - 3430 Tulln
Tel: +43 2272 67133

Austria

Fields of interest
Dairy, water, mycotoxins, PAHs, pesticides, GC, HPLC, spectroscopy, ELISA, analytical chemistry

LEIBETSEDER, Josef Prof., Dr. Dr. h.c.
University of Veterinary Medicine
Department of Nutrition
Veterinärplatz 1
A - 1210 Vienna
Tel :+43 1 250771000
Fax: +43 1 250771090
E-Mail: josef.leibetseder@vu-wien.ac.at

Private address
Argentinier Str. 43
A - 1040 Vienna
Tel: +43 1 5048114

Fields of interest
Dairy, eggs, fish and fish products, meat and meat products, poultry, amino acids, trace elements, mycotoxins, food composition, metabolism, nutrition

LEITNER, Erich Dr.
Graz University of Technology
Department of Biochemistry and Food Chemistry
R & D
Petersgasse 12/2
A - 8010 Graz
Tel :+43 316 873 4522
Fax: +43 316 873 6971
E-Mail: erich.leitner@tugraz.at,
erich@vulgo.com

Fields of interest
Beverages, aroma active compounds, Maillard reaction products, pesticides, GC, MS, sensory analysis, analytical chemistry, packaging, MAP

LUF, Wolfgang Ao.Univ.Prof., Dr.
Veterinary University Vienna
Department of Milk Hygiene, Milk Technology and Food Science
Univ. Professor
Veterinärplatz 1
A - 1210 Vienna
Tel :+43 1 250773501
Fax: +43 1 250773590
E-Mail: wolfgang.luf@vu-wien.ac.at

Private address
Rochnusgasse 16
A - 3002 Purkersdorf
Tel: +43 2231 65464, +43 664 4537115
Fax: 43 2231 65464

Fields of interest
Dairy, eggs, meat and meat products, poultry, oils and fats, water, additives, amino acids, antioxidants, biopolymers, lipids and fatty acids, Maillard reaction products, oxidation reactions, polyphenols, proteins, biogenic amines, mycotoxins, chemometrics, electrochemistry, GC, HPLC, MS, PCR, spectroscopy, adulteration, analytical chemistry, chemical reactions, food composition, nutrition, quality, hygiene, minimal processing, quality assurance, HACCP, storage, thermal processing

MURKOVIC, Michael Dr.
Graz University of Technology
Department of Biochemistry and Food Chemistry
Petersgasse 12/2
A - 8010 Graz
Tel :+43 316 873 6495
Fax: +43 316 873 6971
E-Mail: michael.murkovic@tugraz.at

Fields of interest
Meat and meat products, poultry, oils and fats, antioxidants, vitamins, heterocyclic aromatic amines, HPLC, MS, analytical chemistry, bioavailability, food composition

PFANNHAUSER, Werner Prof., Dr.
Graz University of Technology
Department of Biochemistry and Food Chemistry
Petersgasse 12/2
A - 8010 Graz
Tel :+43 316 873 6471
Fax: +43 316 873 6971
E-Mail: werner.pfannhauser@tugraz.at

Private address
Großfelgitschberg 33
A - 8081 Heiligenkreuz am Waasen

Fields of interest
Beverages, fruits and vegetables, meat and meat products, poultry, oils and fats, spirits, transgenic food, antioxidants, minerals, polyphenols, trace elements, biogenic amines, heterocyclic aromatic amines, PAHs, pesticides, toxic trace elements, AAS, GC, HPLC, sensory analysis, analytical chemistry

SCHMID, Erich R. Prof., Dr.
University of Vienna
Department of Analytical Chemistry
Univ. Prof.
Währingerstr. 38
A - 1090 Vienna
Tel :+49 1 427752302
Fax: +43 1 42779523

Private address
Peter Jordan-Straße 39
A - 1190 Wien
Tel: +43 1 3696820
Fax: +43 1 3696820

Fields of interest
Water, hormones, PAHs, PCBs, pesticides, GC, HPLC, MS, adulteration, analytical chemistry

SIEGMUND, Barbara Dipl.-Ing., Dr.
Graz University of Technology
Institute for Bio- and Food Chemistry
Research Assistant
Petersgasse 12/2
A - 8010 Graz
Tel :+43 316 8736472
Fax: ++43 316 8736971
E-Mail: barbara.siegmund@tugraz.at

Fields of interest
Beverages, fruits and vegetables, aroma active compounds, GC, MS, sensory analysis, analytical chemistry, quality

SONTAG, Gerhard Ao. Prof., Dr.
University of Vienna
Institute for Analytical Chemistry
Teacher
Währinger Str. 38
A - 1090 Vienna
Tel :+43 1 427752303
Fax: +43 1 42779523
E-Mail: gerhard.sontag@univie.ac.at

Private address
Oberlaaer Str. 194
A - 1100 Vienna
Tel: +43 1 6886695

Fields of interest
Beverages, meat and meat products, poultry, heterocyclic aromatic amines, hormones, electrochemistry, HPLC, sensory analysis, adulteration, analytical chemistry

STEINER, Ingrid Ao. Univ.Prof., Dipl.-Ing., Dr.
Technical University Vienna
Institute of Food Chemistry and Food Technology
Getreidemarkt 9
A - 1060 Vienna
Tel :+43 1 5880116002, 5880116011
Fax: +43 1 5880116099
E-Mail: isteiner@mailzserv.tuwien.ac.at

Private address
Heinrich-Kneissl-Gasse 98
A - 1140 Vienna
Tel: +43 1 4190116
Fax: +43 1 4190116

Fields of interest
Beverages, fruits and vegetables, water, additives, antimicrobials, antioxidants, biogenic amines, mycotoxins, AAS, GC, HPLC, MS, TLC, analytical chemistry, microbiology, quality, hygiene, packaging, MAP, preservation, quality assurance, HACCP

TIEFENBACHER, Karl Dr.
Franz Haas Waffelmaschinen AG
Director R&D
Haas-Straße
A - 2100 Leobendorf
Tel: +43 2262 600260
Fax: +43 2262 66455
E-Mail: kt@haaswaffel.at

Private address
Geblergasse 78
A - 1170 Vienna
Tel: +43 1 450163725
Fax: +43 1 450163750

Fields of interest
Bread and cereals, starch, carbohydrates, proteins, PAHs, spectroscopy, management, quality assurance, HACCP, rheology, thermal processing

ULBERTH, Franz Dr.
Agricultural University
Department of Dairy Research and Bacteriology
Assistant Professor and Reader in Food Chemistry
Gregor Mendel-Straße 33
A - 1180 Vienna
Tel :+43 1 476546 100
Fax: +43 1 3299193

Private address
Lederergasse 29/16
A - 1080 Wien

Fields of interest
Dairy, meat and meat products, poultry, oils and fats, aroma active compounds, carotenoids, lipids and fatty acids, chemometrics, GC, HPLC, MS, adulteration

VAN ECKERT, Renate Dr.
Graz University of Technology
Department of Biochemistry and Food Chemistry
Univ.Ass.
Petersgasse 12/2
A - 8010 Graz
Tel :+43 316 873 6467
Fax: +43 316 873 6971
E-Mail: f548vane@mbox.tu-graz.ac.at

Fields of interest
Bread and cereals, dairy, starch, proteins, CE, electrophoresis, HPLC, adulteration, allergology, analytical chemistry

WIEDNER, Peter Dr.
Lebensmitteluntersuchungsanstalt Kärnten

(Carinthian Institute for Food Analysis and Quality Control)
Director
Lastenstraße 40
A - 9020 Klagenfurt
Tel :+43 463 32130
Fax: +43 463 34174
E-Mail: peter.wiedner@ktn.gv.at

Private address
Latschach 8
A - 9064 Pischelsdorf
Tel: +43 4224 3388

Fields of interest
Beverages, dairy, water, additives, trace elements, biogenic amines, PAHs, analytical chemistry, management, quality

---

WINKLER, Johanna Dipl.-Ing.
Saatzucht Gleisdorf GmbH
Breeder
Am Tieberhof 33
A - 8200 Gleisdorf
Tel :+43 3112 2105
Fax: +43 3112 21055
E-Mail: winkler.szgleisdorf@utanet.at

Private address
St. Johann i.d.H. 15
A - 8295 St. Johann i.d.H.
Tel: +43 3332 65538

Fields of interest
Legumes, oils and fats, lipids and fatty acids, proteins, vitamins

---

WEISZ, Richard
Graz University of Technology
Institute of Bio- and Foodchemistry
Petersgasse 12/2
A - 8010 Graz
Tel :+43 316 8736498
Fax: +43 316 8736971
E-Mail: richard.weisz@tugraz.at

Private address
Klosterwiesgasse 241 Pt
A - 8010 Graz
Tel: +43 316 813486
Fax: +43 316 813486

Fields of interest
PAHs, GC, HPLC, analytical chemistry

Belgium

BALDUCK, Paul
Fevia
Advisor
Kortenberglaan 172
B - 1000 Brussels
Tel: +32 2 7430857
Fax: +32 2 7339426
E-Mail: pb@fevia.be

Private address
Mechelsesteenweg 1105
B - 3020 Winksele
Tel: +32 16 480712
Fax: +32 16 480712

Fields of interest
Additives, allergology, metabolism, nutrition, biotechnology, fermentation, genetic engineering, rheology

---

BEERNAERT, Hedwig Chemist
Scientific Institute of Public Health
Quality Assurance Manager
Juliette Wytsman 14
B - 1050 Brussels
Tel :+32 2 6425186
Fax: +32 2 6425001
E-Mail: h.beernaert@iph.fgov.be

Private address
Kasteelstraat 22
B - 9660 Brakel
Tel: +32 55 422226

Fields of interest
Carbohydrates, lipids and fatty acids, minerals, steroids, trace elements, vitamins, hormones, PAHs, PCBs, pesticides, toxic trace elements, AAS, GC, HPLC, MS, spectroscopy, TLC, analytical chemistry, food composition, management, nutrition, quality, regulative issues, quality assurance, HACCP

DE BLOCK, Jan Ph.D. Biochemistry
Agricultural Research Centre - Ghent
Department of Animal Product Quality
Research Assistant
Brusselsesteenweg 370
B - 9090 Melle
Tel :+32 9 2521861
Fax: +32 9 2525085
E-Mail: dvk@clo.fgov.be

Private address
Hellestraat 20
B - 2530 Boechout
Tel: +32 3 4541231

Fields of interest
Dairy, meat and meat products, poultry, Maillard reaction products, proteins, CE, electrophoresis, HPLC, spectroscopy, adulteration, thermal processing

DE BRABANDER, Hubert Dr., Prof.
University of Gent
Department of Veterinary Food Inspection, Lab of Chemical Analysis
Salisburylaan 133
B - 9820 Merelbeke
Tell: +32 9 2647460
Fax: +32 9 2647492

Private address
J. Heremansstr. 44
B - 9000 Gent
Tel: +32 9 2217311

Fields of interest
Meat and meat products, poultry, steroids, trace elements, drugs, hormones, PCBs, GC, HPLC, MS, TLC, GC -MS, LC-MSn, analytical chemistry

DE RUYCK, Hendrik Engineer
Goverment Research Station of Dairying
CLO-DVK
Assistant
Brusselsesteenweg 370
B - 9090 Melle
Tel :+32 9 2521861
Fax: +32 9 2525085
E-Mail: h.deruyck@clo.fgov.be

Private address
Nazarethsesteenweg 36
B - 9800 Deinze
Tel: +32 9 3801418
Fax: +32 9 3801418

Fields of interest
Dairy, eggs, meat and meat products, poultry, drugs, HPLC, MS, analytical chemistry, food composition, quality, regulative issues

DEELSTRA, Hendrik Ph.D.
University Antwerpen (U.I.A.)
Professor
Univ. Plein 1
B - 2610 Wilryk
Tel :+32 3 8202715
Fax: +32 3 8202734
E-Mail: labrow@via.ua.ac.be

Private address
Graaf Witgerstraat, 8
B - 2640 Mortsel
Tel: +32 3 4408003

Fields of interest
Minerals, trace elements, toxic trace elements, AAS, analytical chemistry, bioavailability, food composition, nutrition, infants foods, total diets

DEWEGHE, Liane Ing.
Experimental & Analytical Station - COOVI
Laboratory Manager
E. Gryzonlaan, 1
B - 1070 Brussels
Tel :+32 2 5267250
Fax: +32 2 5267259

Private address
Ninoofsesteenweg 252
B - 1700 Dilbeek
Tel: +32 2 5679863

Fields of interest
Food Products, carbohydrates, dietary fibres, lipids and fatty acids, minerals, proteins, AAS, GC, HPLC, food composition, microbiology

DIRINCK, Patrick Ph.D.
Catholic Technical Univ. of East-Flanders
Chemical and Biochemical Research Centre (CBOK)
Professor, Research Coordinator
Gebr. Desmetstraat 1
B - 9000 Gent
Tel :+32 9 2658638
Fax: +32 9 2658638

Private address
Graaf van Landaststraat 152
B - 9700 Oudenaarde
Tel: +32 55 315059

Fields of interest
Beer, beverages, coffee, tea, cocoa, dairy, fruits and vegetables, oils and fats, spices, spirits, starch, wine, aroma active compounds, chemometrics, GC, HPLC, MS, sensory analysis, packaging, MAP

HERMAN, Lieve Ph.D. in Biological Science
Goverment Dairy Research Station
Centre of Agricultural Research
Scientific colaborator
Brusselsesteenweg 370
B - 9090 Melle
Tel :+32 9 2521861
Fax: +32 9 2525085
E-Mail: l.herman@clo.fgov.be

Fields of interest
Dairy, eggs, meat and meat products, poultry, transgenic foods, microbiology, quality, biotechnology, quality assurance, HACCP

Belgium

ILSBROUX, Ingrid Ph.D.
Industriele Hogeschool Group
Vesaliusstraat 13
B - 3000 Leuven
Tel :+32 16 301030, 301092
Fax: +32 16 301040
E-Mail: ingrid.ilsbroux@groept.be

Private address
Bovenbosstraat 108
B - 3053 Haasrode

Fields of interest
Beer, bread and cereals, dairy, oils and fats, starch, sweeteners, transgenic foods, microbiology, biotechnology, fermentation

---

KNOWLES, Michael Ernest B.Pharm., Ph.D., C.Chem., FRSC, FIFST
Coca-Cola Greater Europe
Director, Scientific and Regulatory Affairs
Chaussee de Mons 1424
B - 1070 Brussels
Tel :+32 2 5291710
Fax: +32 2 5291718
E-Mail: mknowles@eur.ko.com

Private address
Avenue du Faucon, 28
B - 1640 Rhode - St. Genese
Tel: +32 2 3585130
Fax: +32 2 3585130

Fields of interest
Beverages, additives, polyphenols, plant toxins, sensor technology, chemical reactions, consumer research, nutrition, toxicology, packaging, MAP

OOGHE, Wilfried Dr., Prof.
University of Gent, Laboratorium voor Bromatologie
Professor
Harelbekestr. 72
B 9000 Gent
Tel :+32 9 2648127
Fax: +32 9 2648299
E-Mail: wilfried.ooghe@rug.ac.be

Private address
Kollebloemstr. 63
B - 9030 Gent
Tel: +32 9 2268580
Fax: +32 9 2268580

Fields of interest
Beverages, fruits and vegetables, meat and meat products, poultry, wine, amino acids, polyphenols, proteins, AAS, HPLC, spectroscopy, adulteration, analytical chemistry, consumer research, food composition, nutrition, quality

---

SCHREYEN, Luc Dr. IR.
N.V. INEX Dairy Industries
Production Manager
Meulestraat 19
B - 9520 Bavegem
Tel :+32 9 3638282
Fax: ++32 9 3622037

Private address
Scheurbroek 6
B - 9860 Oosterzele
Tel: +32 9 3627480

Fields of interest
Dairy, PCBs, GC, MS, sensory analysis, adulteration, quality, fermentation, quality assurance, thermal processing

---

STEEGMANS, Monique Lic. Chemistry
Tiense Suikerraffinardeky
Raffinerie Tirlemontoise/Orafti
Adjunct head analytical lab
Aandorenstraat 1
B - 3300 Tienen
Tel: +32 16 801253
Fax: +32 16 820317

Private address
Meeuwerbaan 93
B - 3990 Peer
Tel: +32 11 793561

Fields of interest
Sugar, honey, sugar alcohols, sweeteners, carbohydrates, dietary fibres, fat replacers, GC, HPLC, spectroscopy, analytical chemistry, food composition

---

TEMMERMAN, Guy Lic.
Ministry of Public Health
Food Inspection
RAC-Esplanade, Pachecolaan 19 B 5
B - 1010 Brussels
Tel: +32 2 2104843
Fax: +32 2 2104816
E-Mail: guy.temmerman@health.fgov.be

Private address
Hof Leeuwergem 37
B - 9300 Aalst
Tel: +32 53 788716
Fax: +32 53 788716

Fields of interest
Regulative issues

VAN PETEGHEM, Carlos Ph.D., Pharmacist
Gent University
Full Professor of Food Chemistry and Toxicology

Harelbekestraat 72
B - 9000 Gent
Tel: +32 9 2648134
Fax: +32 9 2648115
E-Mail: carlos.vanpeteghem@rug.ac.be

Private address
Evergemsesteenweg 69
B - 9032 Gent
Tel: +32 9 2539596
Fax: +32 9 2539596

Fields of interest
Meat and meat products, poultry, steroids, hormones, mycotoxins, PCBs, GC, HPLC, MS, nutrition, toxicology

---

VAN RENTERGHEM, Roland Lic.
DVK-CLO
Head of section
Brusselsesteenweg 370
B - 9090 Melle
Tel: +32 9 2521861
Fax: +32 9 2525085
E-Mail: r.vanrenterghem@clo.fgov.be

Private address
J. Destreelaan 52
B - 9050 Gent

Fields of interest
Dairy, mycotoxins, PCBs, AAS, GC, HPLC, spectroscopy, adulteration, analytical chemistry, food composition

---

VAN VYNCHT, Gery Ph.D.
Institute for Reference Materials and Measurements (I.R.M.M.) - JRC - GEEL
Researcher - Analytical Chemistry Department
Retiesenweg
B - 2440 Geel
Tel: +32 14 571339
Fax: +32 14 584273
E-Mail: van-vyncht@irmm.jrc.be

Private address
Rue Arzieres 3
B - 6440 Froidchapelle
Tel: +32 60 412236

Fields of interest
Meat and meat products, poultry, oils and fats, water, antimicrobials, proteins, steroids, trace elements, vitamins, drugs, hormones, PAHs, PCBs, pesticides, plant toxins, toxic trace elements, GC, HPLC, MS, TLC, adulteration, analytical chemistry, food composition, metabolism, quality

---

VANLUCHENE, Eric Dr. Chem.
Ana Bio Tec
Research and Development
Reibroekstraat 13
B - 9940 Evergem
Tel: +32 9 2536030
Fax: +32 9 2535537
E-Mail: vanluchene.lazeure@village.uunet.be

Private address
Parelhoenstraat 27
B - 9000 Gent
Tel: +32 9 2211273
Fax: +32 9 2211273

Fields of interest
Lipids and fatty acids, steroids, vitamins, mycotoxins, PCBs, pesticides, CE, GC, HPLC, MS

Bulgaria

DALEV, Pencho Dr., Assoc. Professor, Ph.D., D.Sc.
University of Sofia
Faculty of Biology
Head of Laboratory for New Protein Sources
Bul. Dragan Tsankov 8
BG - 1421 Sofia
Tel: +359 2 656291
Fax: +359 2 658079

Private address
kv. Hypodruma, bl. 23 A
BG - 1612 Sofia
Tel: +359 2 543805

Fields of interest
Legumes, meat and meat products, poultry, amino acids, biopolymers, dietary fibres, proteins, plant toxins, HPLC, enzymology, food composition, nutrition, biotechnology

---

DONCHEVA, Ivanka Dipl. Chemist, Ph.D.
National Centre of Hygiene, Medical Ecology and Nutrition
Research Associate
D. Nestorov 15
BG - 1431 Sofia
Tel: +359 2 5812247
Fax: +3592 598148

Private address
Udovo 10/V/45
BG - 1463 Sofia
Tel: +359 2 547726

Fields of interest
Bread and cereals, coffee, tea, cocoa, legumes, spices, mycotoxins, HPLC, TLC, analytical chemistry,

nutrition, quality assurance, HACCP

IVANOV, Kalintcho Ph.D., Associate Professor
Higher Institute of Food and Flavour Technology
Assistant Professor
26 Maritza blvd
BG - 4002 Plovdiv
Tel: +359 32 44181226
Fax: +359 32 440102

Private address
2 Svilen Russev
BG - 4000 Plovidv
Tel: +359 2 433801

Fields of interest
Cosmetics, oils and fats, additives, antioxidants, aroma active compounds, carbohydrates, AAS, spectroscopy, analytical chemistry, food composition

KOVATCHEVA-APOSTOLOVA, Elena Ph.D.
Higher Institute for Food and Flavour Industries
Associate Professor
Maritza 26
BG - 4002 Plovdiv
Tel: +359 32 44181283
Fax: +359 32 440101

Private address
Jagodastr. 1
BG - 4004 Plovdiv
Tel: +359 32 777096

Fields of interest
Additives, antioxidants, vitamins, GC, HPLC, spectroscopy, analytical chemistry, food composition, quality

NEICHEVA, Anastasia Ph.D., Assitant Professor
Higher Institute of Food and Flavour Technology
Assistant Professor
26 Maritza blvd
BG - 4002 Plovdiv
Tel: +359 32 44181226
Fax: +359 32 440102
E-Mail: aneicheva@yahoo.com

Private address
48 Bratia Buckstone
BG - 4000 Plovidv
Tel: +359 2 762881

Fields of interest
Fruits and vegetables, additives, antioxidants, pesticides, GC, HPLC, MS, TLC, analytical chemistry, metabolism

OBRETONOV, Tzvetan D.Sc.
Higher Institute of Food and Flavour Industry of Bulgaria
Professor
Maritza bd. 26
BG - 4000 Plovdiv
Tel: +359 32 44181281
Fax: +359 32 440102
E-Mail: obreten@hiffi.plovdiv.acad.bg

Private address
Milin Kamak 10
BG - 4002 Plovdiv
Tel: +359 32 432628

Fields of interest
Antioxidants, aroma active compounds, colours, Maillard reaction products, GC, MS, spectroscopy, thermal anlaysis, chemical reactions, thermal processing

POPOV, Dimitre Ph.D., Associate Professor
Higher Institute of Food and Flavour Technology
Head of Department
26 Maritza blvd
BG - 4002 Plovdiv
Tel: +359 32 44181313
Fax: +359 32 440102

Private address
18 P. Shilev, ap. 38
BG - 4003 Plovidv
Tel: +359 32 566408

Fields of interest
Wine, minerals, electrochemistry, analytical chemistry

RIBAROVA, Fanny Ph.D., Ass. Prof.
National Centre of Hygiene
Department of Food Chemistry
Head of Food Composition Lab
D. Nestorov 15
BG - 1431 Sofia
Tel: +359 2 591011, 5812238
Fax: +359 2 598148

Private address
Solunska 29
BG - 1000 Sofia
Tel: +359 2 872861

Fields of interest
Amino acids, antioxidants, carotenoids, lipids and fatty acids, vitamins, HPLC, food composition, metabolism, nutrition

RIZOV, Nicolay Ph.D.
National Center of Hygiene, Medical Ecology and Nutrition

Director
D. Nestorov 15
BG - 1431 Sofia
Tel: +359 2 9581894
Fax: +359 2 9581277
E-Mail: n.rizov@nch.aster.net

Private address
Chercovna 68
BG - 1505 Sofia
Tel: +359 2 437844

Fields of interest
Additives, amino acids, antioxidants, aroma active compounds, carbohydrates, colours, ethanol, lipids and fatty acids, minerals, proteins, trace elements, vitamins, PAHs, PCBs, pesticides, toxic trace elements, AAS, GC, HPLC, MS, spectroscopy, TLC, analytical chemistry, management, nutrition, quality, regulative issues, toxicology, HACCP

STANOEVA, Elena Assoc. Univ. Prof., Ph.D.
University of Sofia
Department of Organic Chemistry
James Bourchier Avenue 1
BG - 1126 Sofia
Tel: +359 2 6256225
Fax: +359 2 9625438
E-Mail: estanoeva@chem.uni-sofia.bg

Private address
Svetlostrui str. 8v
BG - 1111 Sofia
Tel: +359 2 710216

Fields of interest
Additives, aroma active compounds, alkaloids, biogenic amines, heterocyclic aromatic amines, pesticides, plant toxins, MS, TLC, chemical reactions, metabolism, regulative issues

---

TOPALOVA, Ivanka Chrstova Dr.
University of Sofia 'St. Kliment Ohridsky'
Faculty of Chemistry
Associate Professor
J. Bourchier Bul. 1
BG - Sofia 1126
Tel: +359 2 6256367
Fax: +359 2 9625438
E-Mail: Itopalova@chem.uni-sofia.bg

Private address
Busludscha 39
BG - 1463 Sofia
Tel: +359 2 518340

Fields of interest
Cosmetics, water, wine, additives, aroma active compounds, ethanol, pesticides, GC, HPLC

TSANEV, Roumen Chem.Eng., Ph.D.
National Centre of Hygiene
Research Associate
D. Nestorov 15
BG - 1431 Sofia
Tel: +359 2 5812243, 406
Fax: +359 2 598148

Private address
Udovo 10/V/45
BG - 1463 Sofia
Tel: +359 2 547726

Fields of interest
Oils and fats, fat replacers, lipids and fatty acids, oxidation reactions, GC, analytical chemistry, food composition, nutrition, hygiene

---

YANISHLIEVA-MASLAROVA, Nedjalka Professor, Dr.Sc.
Bulgarian Academy of Sciences
Institute of Organic Chemistry
Head of Dept. of Lipid Chemistry
Acad. G. Bonchev, Bl.9
BG - 1113 Sofia
Tel: +359 2 9796696, 7334178, 7334173
Fax: +359 2 700225
E-Mail: nelly@orgchm.ba.bg

Private address
kv. Mladost, Bl. 95/2
BG - 1756 Sofia
Tel: +359 2 719769

Fields of interest
Oils and fats, spices, antioxidants, carbohydrates, lipids and fatty acids, oxidation reactions, polyphenols, GC, TLC, food composition

Croatia

BAZULIC, Davorin Dr.Sc.
Croatian Veterinary Institute, Department for Residue Analysis
Associate Professor, Head of Department for Residue Analysis
Savska C. 143
HR - 10000 Zagreb
Tel: +358 1 6123601
Fax: +358 1 6190841

Private address
Mlinovi 18
HR - 10000 Zagreb
Tel: +385 1 4673245
Fax: +385 1 4673245

Fields of interest
Dairy, eggs, fish and fish products, meat and meat products, poultry, drugs, PCBs, pesticides, GC,

HPLC, nutrition, toxicology, quality assurance, HACCP

GALIC, Kata Prof.
Faculty of Food Technology and Biotechnology
Professor
Pierottijeva 6
HR - 10000 Zagreb
Tel: +385 1 4605002
Fax: +385 1 4836083
E-Mail: kgalic@pbf.hr

Private address
D. Cesaric 15
HR - 10090 Zagreb
Tel: +385 1 3791389

Fields of interest
Biotechnology, packaging, MAP, storage

GOJMERAC, Tihomira Ph.D.
Croatian Veterinary Institute
Head, Department of Biochemistry
Savska C. 143
HR - 10000 Zagreb
Tel: +385 1 6123666
Fax: +385 1 6190841
E-Mail: Tihomira.gojmerac@mari.veinst.hr

Private address
Vrhovac 40
HR - 10000 Zagreb
Tel: +385 1 3703793

Fields of interest
Fish and fish products, meat and meat products, poultry, water, amino acids, lipids and fatty acids, proteins, steroids, vitamins, hormones, mycotoxins, pesticides, HPLC, MS, spectroscopy, TLC, analytical chemistry, toxicology, high pressure technology, quality assurance, HACCP

HARDI, Jovica Dr.Sc.
University J.J. Strossmayer Osijek
Faculty of Food Technology
Associate Professor
F. Kuhaca 18, P.O.Box 709
HR - 31000 Osijek
Tel: +385 31 224310
Fax: +385 31 207115
E-Mail: jovica.hardi@ptfos.hr

Private address
Kapucinska 26
HR - 31000 Osijek
Tel: +385 31 201128

Fields of interest
Dairy, amino acids, aroma active compounds, lipids and fatty acids, pesticides, GC, MS, food

composition, biotechnology, fermentation

HASENAY, Damir M.Sc.
University J.J. Strossmayer Osijek
Faculty of Food Technology
Assistant
Kuhaceva 18
HR - 31107 Osijek
Tel: +385 31 224300
Fax: +385 31 207115
E-Mail: hasenayd@ptfos.hr

Private address
J.Huttlera 23
HR - 31000 Osijek
Tel: +385 31 506968

Fields of interest
AAS, electrochemistry, polarography, packaging, MAP

HEGEDUSIC, Vesna Dr. sc.
University of Zagreb
Faculty of Food Technology and Biotechnology
Professor, Vice dean
Pierrotijeva 6
HR - 10000 Zagreb
Tel: +385 1 4605287
Fax: +385 1 4836083
E-Mail: vheged@mapbf.pbf.hr

Private address
Zajceva 6
HR - 10000 Zagreb
Tel: +385 1 3631302

Fields of interest
Fruits and vegetables, thermal anlaysis, drying, freezing, rheology, thermal processing

HORVATIC, Marija Dr.Sc.
University of Zagreb
Faculty of Pharmacy and Biochemistry, Department of Food Chemistry
Univ.Prof.
A. Kovacica 1
HR - 10000 Zagreb
Tel: +385 1 4612690
Fax: +385 1 4612691

Private address
Zeleni trg 1
HR - 10000 Zagreb
Tel: +385 1 6199499

Fields of interest
Bread and cereals, legumes, amino acids, proteins, trace elements, bioavailability, food composition,

nutrition, quality

HRUSKAR, Mirjana M.Sc.
Faculty of Food Technology and Biotechnology
Assistant
Pierottijeva 6
HR - 10000 Zagreb
Tel: +385 1 4605043
Fax: +385 1 4836083
E-Mail: mhruskar@mapbf.pbf.hr

Private address
Paljetkova 10
HR - 10000 Zagreb
Tel: +385 1 3882049

Fields of interest
Sensory analysis, food composition, nutrition, quality, quality assurance, HACCP

KLAPEC, Tomislav M.Sc.
University J.J. Strossmayer Osijek
Faculty of Food Technology
Assistant
F. Kuhaca 18, P.O.Box 709
HR - 31107 Osijek
Tel: +385 31 224364
Fax: +385 31 207115
E-Mail: tomi@ptfos.hr

Private address
Opatijska 35
HR - 31100 Osijek
Tel: +385 31 563190

Fields of interest
Minerals, trace elements, AAS, nutrition

KORDIC, Jasna M.Sc..
University J.J. Strossmayer Osijek
Faculty of Food Technology
Assistant
F. Kuhaca 18, P.O.Box 709
HR - 31000 Osijek
Tel: +385 31 224300, 224357
Fax: +385 31 207115
E-Mail: jkordic@ravnica.ptfos.hr

Private address
Pozeska 46
HR - 31000 Osijek
Tel: +385 31 580513

Fields of interest
Meat and meat products, poultry, nutrition, food technology

KOVAC, Spomenka Dr. sc.
University J.J. Strossmayer Osijek
Faculty of Food Technology
Professor
Kuhaceva 18
HR - 31107 Osijek
Tel: +385 31 224300
Fax: +385 31 207115
E-Mail: kps@os.tel.hr

Private address
V.I. Mestrovica 8
HR - 31000 Osijek
Tel: +385 31 203485
Fax: +385 31 203485

Fields of interest
Sweeteners, amino acids, pesticides, toxic trace elements, TLC, nutrition, biotechnology

MANDIC, Milena L. Prof., Dr. sc.
University J.J. Strossmayer Osijek
Faculty of Food Technology
Vice Dean, Head of the Department of Nutrition and Food quality control
F. Kuhaca 18, P.O.Box 709
HR - 31107 Osijek
Tel: +385 31 224300
Fax: +385 31 207115
E-Mail: milena.mandic@ptfos.hr

Private address
A.M. Reljkovica
HR - 31000 Osijek
Tel: +385 31 208116

Fields of interest
Dietary Fibres, minerals, trace elements, vitamins, trace elements, vitamins, toxic trace elements, AAS, sensory analysis, food composition, nutrition

NOVAKOVIC, Predrag Prof. Dr. sc.
University J.J. Strossmayer Osijek
Faculty of Food Technology
Professor
F. Kuhaca 18, P.O.Box 709
HR - 31000 Osijek
Tel: +385 31 224300
Fax: +385 31 207115
E-Mail: predrag.novakovic.ptfos.hr

Private address
W. Wilsona 17
HR - 31000 Osijek
Tel: +385 31 570740
Fax: +385 31 570 740

Fields of interest
Dairy, meat and meat products, poultry, dietary fibres, fat replacers, lipids and fatty acids, food composition, metabolism, nutrition, fermentation, thermal processing

Croatia

PILIZOTA, Vlasta Prof. Dr. sc.
University J.J. Strossmayer Osijek
Faculty of Food Technology
Professor
F. Kuhaca 18, P.O.Box 709
HR - 31000 Osijek
Tel: +385 31 224300, 224311
Fax: +385 31 207115
E-Mail: vlasta.pilizota@ptfos.hr

Private address
Sjenjak 19
HR - 31000 Osijek
Tel: +385 31 574524

Fields of interest
Fruits and vegetables, colours, Maillard reaction products, polyphenols, GC, thermal anlaysis, food composition, minimal processing, packaging, MAP, preservation

---

POSPISIL, Jasna Dr.Sc.
Faculty of Food Technology and Biotechnology
Univ. Prof.
Pierottijeva 6
HR - 10000 Zagreb
Tel: +385 1 4605168
Fax: +385 1 4605072
E-Mail: jpospi@mapbf.pbf.hr

Private address
II Crikvenicka 11
HR - 10000 Zagreb
Tel: +385 1 324351
Fax: +385 1 317897

Fields of interest
Fruits and vegetables, legumes, antioxidants, colours, dietary fibres, HPLC, adulteration, nutrition, minimal processing, packaging, MAP

POZDEROVIC, Andrija Professor Dr. sc.
University J.J. Strossmayer Osijek
Faculty of Food Technology
Professor
Kuhaceva 18
HR - 31000 Osijek
Tel: +385 31 224300, 224313
Fax: +385 31 207115
E-Mail: andrija.pozderovic@ptfos.hr

Private address
N. Subic Zrimskog 27
HR - 31000 Osijek
Tel: +385 31 573613

Fields of interest
Fruits and vegetables, potatoes, wine, aroma active compounds, colours, GC, MS, sensory analysis, filtration, freezing, rheology

---

PRIMORAC, Ljiljana Dr.Sc.
University J.J. Strossmayer Osijek
Faculty of Food Technology
Assistant
Kuhaceva 18, p.p. 709
HR - 31107 Osijek
Tel: +385 31 224300
Fax: +385 31 207115
E-Mail: ljiljana.primorac@ptfos.hr

Private address
F.Krezme 9
HR - 31000 Osijek
Tel: +385 31 551958

Fields of interest
Lipids and fatty acids, trace elements, GC, food composition, nutrition

---

SAPUNAR-POSTRUZNIK, Jasenka Dr.Sc.
Croatian Veterinary Institute, Department for Residue Analysis
Associate Professor, Head of Laboratory for AAS. Anal.
Savska C. 143
HR - 10000 Zagreb
Tel: +358 1 6123601
Fax: +358 1 6190841

Private address
Gunduliceva 36
HR - 10000 Zagreb
Tel: +385 1 4856036

Fields of interest
Dairy, eggs, fish and fish products, meat and meat products, poultry, minerals, trace elements, toxic trace elements, AAS, nutrition, toxicology

---

SEBECIC, Blazenka Dr. sc.
University of Zagreb
Faculty of Pharmacy and Biochemistry, Department of Food Chemistry
Univ. Prof.
A. Kovacica 1
HR - 10000 Zagreb
Tel: +385 1 4612690
Fax: +385 1 4612691

Private address
IV Cvjetno Naselje 14
HR - 10000 Zagreb
Tel: +385 1 6196114

Fields of interest
Bread and cereals, starch, minerals, proteins, vitamins, toxic trace elements, AAS, spectroscopy, nutrition, rheology

SUBARIC, Drago Dr. Sc..
University J.J. Strossmayer Osijek
Faculty of Food Technology
Assistant
F. Kuhaca 18, P.O.Box 709
HR - 31000 Osijek
Tel: +385 31 224300
Fax: +385 31 207115
E-Mail: drago.subaric@ptfos.hr

Private address
Sjenjak 34
HR - 31000 Osijek
Tel: +385 31 573620

Fields of interest
Fruits and vegetables, sugar, honey, sugar alcohols, colours, polyphenols, GC, minimal processing, rheology

UGARCIC-HARDI, Zaneta Dr.Sc.techn. ETH
University J.J. Strossmayer Osijek
Faculty of Food Technology
Associate Professor
F. Kuhaca 18, P.O.Box 709
HR - 31000 Osijek
Tel: +385 31 224308
Fax: +385 31 207115
E-Mail: zaneta.ugarcic-hardi@ptfos.hr

Private address
Kapucinska 26
HR - 31000 Osijek
Tel: +385 31 201128

Fields of interest
Bread and cereals, additives, carotenoids, dietary fibres, minerals, electrophoresis, sensory analysis, food composition, quality, fermentation, rheology, storage

VAHCIC, Nada Ph.D.
Faculty of Food Technology and Biotechnology
Associate Professor
Pierottijeva 6
HR - 10000 Zagreb
Tel: +385 1 4605048
Fax: +385 1 4836083
E-Mail: nvahcic@mapbf.pbf.hr

Private address
Slavenskog 23
HR - 10000 Zagreb
Tel: +385 1 3881463

Fields of interest
Sensory analysis, food composition, quality, regulative issues, quality assurance, HACCP

VEDRINA-DRAGOJEVIC, Irena Dr. sc.
University of Zagreb
Department of Food Chemistry, Faculty of Pharmacy and Biochemistry
Associate Professor
A. Kovacica 1
HR - 10000 Zagreb
Tel: +385 1 4612690
Fax: +385 1 4612691

Private address
Ruzmarinka 21
HR - 10000 Zagreb
Tel: +385 1 228339

Fields of interest
Nutritional biochemistry

Czech Republic

BENDA, Vladimir M.V.D., Ph.D., D.Sc., Assoc. Prof.
Institute of Chemical Technology
Department of Biochemistry and Microbiology
Technická 3
CZ - 16628 Praha 6
Tel: +42 2 24353697
Fax: +42 2 3114769
E-Mail: vladimir.benda@vscht.cz

Private address
Ruzyne 587
CZ - 16100 Praha
Tel: +42 2 3010301

Fields of interest
Meat and meat products, poultry, transgenic food, proteins, pesticides, electrophoresis, consumer research, immunology, microbiology, genetic engineering

BUBNIK, Zdenek Ph.D.
Institute of Chemical Technology Prague
Associate Professor
Technická 1905
CZ - 16628 Praha 6
Tel: +42 2 2435 3112
Fax: +42 2 3119990
E-Mail: zdenek.bubnik@vscht.cz

Private address
Fortova 67
CZ - 18100 Praha 8
Tel: +42 2 8541842

Fields of interest
Sugar, honey, sugar alcohols, carbohydrates, HPLC, consumer research, food composition, quality, biotechnology, filtration, crystallisation, membrane separation

Czech Republic

CEJPEK, Karel M.Sc., Ph.D.
Institute of Chemical Technology
Assistent Professor
Technicka 5
CZ - 16628 Praha
Tel: +420 2 24353178
Fax: +420 2 3119990
E-Mail: karel.cejpek@vscht.cz

Fields of interest
Fruits and vegetables, Maillard reaction products,isothiocyanates, GC, HPLC, analytical chemistry, food composition, nutrition

CHUMCHALOVÁ, Jana Ph.D.
The Prague Institute of Chemical Technology
Assistant
Technická 3
CZ - 16628 Praha
Tel: +420 2 24353275
Fax: +420 2 3119990
E-Mail: jana.chumchalova@vscht.cz

Private address
Hrebecská 2648
CZ - 27100 Jkladno

Fields of interest
Dairy, antimicrobials, mycotoxins, electrophoresis, HPLC, PCR, microbiology, nutrition, biotechnology, fermentation, genetic engineering, hygiene, preservation

COPÍKOVÁ, Jana Ing., C.Sc.
Institute of Chemical Technology
Assistant Professor
Technická 5
CZ - 16628 Praha 6
Tel: +42 2 24353114
Fax: +42 2 3119990

Private address
Mimonská 632
CZ - 19000 Praha 9
Tel: +42 2 883430

Fields of interest
Coffee, tea, cocoa, carbohydrates, lipids and fatty acids, GC, HPLC, spectroscopy, analytical chemistry

CULIK, Jirí Ph.D.
Research Institute of Brewing and Malting
Analytical Division
Deputy Head of Analytical Division
Lípová 15
CZ - 12044 Praha 2
Tel: +42 02 24915384
Fax: +42 02 295037

Private address
Ke Klimentce 1/1852
CZ - 15000 Praha 5
Tel: +42 2 57211260

Fields of interest
Beer, beverages, meat and meat products, poultry, aroma active compounds, PAHs, PCBs, N-nitrosamines, GC, HPLC, analytical chemistry

CURDA, Ladislav Ph.D.
The Prague Institute of Chemical Technology
Assistant Professor
Technická 3
CZ - 16628 Praha
Tel: +420 2 24353831
Fax: +420 2 3119990
E-Mail: ladislav.curda@vscht.cz

Private address
Evropská 201
CZ - 16100 Prague
Tel: +42 2 3166638

Fields of interest
Dairy, minerals, proteins, CE, chemometrics, spectroscopy, quality, biotechnology, rheology

DAVIDEK, Jiri Prof., Ing., Dr.Sc.
Institute of Chemical Technology
Department of Food Chemistry and Analysis
Professor
Technická 1905
CZ - 16628 Praha 6
Tel: +42 02 24353177
Fax: +42 02 3119990

Private address
Obrovskeho 12/317
CZ - 14100 Praha 4
Tel: +42 2 72764777

Fields of interest
Additives, aroma active compounds, Maillard reaction products, vitamins, plant toxins, GC, HPLC, analytical chemistry, chemical reactions, quality

DEMNEROVA, Katerina Ph.D., Prof.
Institute of Chemical Technology
Department of Biochemistry and Microbiology
Head of Microbial Division
Technická 5
CZ - 16628 Praha 6
Tel: +42 2 2435 3075, 5172
Fax: +42 2 2435 3075

Private address
Milady Horakove 63
CZ - 17000 Praha 7

Tel: +42 2 33373950

DOLEZAL, Marek M.Sc., Ph.D.
Institute of Chemical Technology
Assistent Professor
Technická 5
CZ - 16628 Praha 6
Tel: +420 2 24353179
Fax: +420 2 3119990
E-Mail: marek.dolezal@vscht.cz

Fields of interest
Dairy, eggs, oils and fats, aroma active compounds, aroma active compounds, lipids and fatty acids, GC, analytical chemistry, chemical reactions, food composition, nutrition

DOSTÁLOVÁ, Jana Ph.D., Dipl.-Ing.
Institute of Chemical Technology
Department of Food Chemistry and Analysis
Teacher - Associate Professor
Technická 5
CZ - 16628 Praha 6
Tel: +420 2 24353264
Fax: +420 2 3119990
E-Mail: dostaloj@vscht.cz

Private address
Voronezská 4
C - 10100 Praha 10
Tel: +42 2 71743140

Fields of interest
Legumes, sweeteners, carbohydrates, lipids and fatty acids, analytical chemistry, chemical reactions, consumer research, nutrition, quality, thermal processing

DRDAK, Milan Univ. Prof., Dr., Dipl.-Ing., Dr.Sc.
Technical University Brno
Faculty of Chemistry
Director Institute for Chemistry of Foodstuffs and Biotechnologies
Purkynova 118
CZ - 61200 Brno
Tel: +425 41 149320
Fax: +425 41 211697
E-Mail: drdak@fch.vutbr.cz

Private address
Bernolakova 2424/ 10B
SK - 90101 Malacky

Fields of interest
Beverages, fruits and vegetables, transgenic foods, wine, colours, sensory analysis, food composition, quality, preservation, thermal processing

FILIP, Vladimir Assoc. Prof., Ph.D., Doz.Dr.
Prague Institute of Chemical Technology
Head of the Department of Dariy and Fat Technology
Technická 3
CZ - 16628 Praha 6
Tel: +420 2 24353268
Fax: +420 2 3119990
E-Mail: vladimierfilip@vscht.cz

Private address
Laudova 1002
CZ - 16300 Praha 6

Fields of interest
Cosmetics, dairy, oils and fats, antimicrobials, lipids and fatty acids, oxidation reactions, GC, HPLC, TLC, chemical reactions, food composition, quality assurance, rheology, fat technology

HOLANDOVÁ, Katerina M.Sc., Ph.D.
Institute of Chemical Technology
Assistant Professor
Technická 3
CZ - 16628 Prahae 6
Tel: +420 2 24353218
Fax: +420 2 24353185
E-Mail: katerina.holadova@vscht.cz

Private address
Karolíny Svetlé 14
CZ - 11000 Prague 1
Tel: +42 2 263127

Fields of interest
Fruits and vegetables, oils and fats, water, mycotoxins, PCBs, pesticides, GC, HPLC, MS, analytical chemistry, food composition

HOLASOVA, Marie M.Sc.
Food Research Institute Prague
Research Worker
Radiová 7
CZ - 10231 Praha 10
Tel: +42 2 702331
Fax: +42 2 701983
E-Mail: m.holasova@vupp.cz

Private address
Na Petrinách 29
CZ - 16200 Praha 6
Tel: +42 2 365934
Fax: +42 2 3165526

Fields of interest
Antioxidants, vitamins, HPLC, analytical chemistry, food composition, nutrition, quality

HRNCIRIK, Karel M.Sc., Ph.D.
Institute of Chemical Technology

Assistent Professor
Technicka 5
CZ - 16628 Praha
Tel: +420 2 24353179
Fax: +420 2 3119990
E-Mail: hrncirik@vscht.cz

Fields of interest
Fruits and vegetables, vitamins, glucosinolates, plant toxins, GC, HPLC, analytical chemistry, food composition, nutrition, toxicology

INGR, Ivo Dipl.-Ing, Dr.Sc., Professor
Mendel University of Agriculture and Forestry
Department of Food Technology
Professor of Food Technology
Zemedelska 1
CZ - 61300 Brno
Tel: +42 5 45133197
Fax: +42 5 45212044
E-Mail: zlato@mendelu.cz

Private address
Milenova 3
CZ - 63800 Brno
Tel: +42 5 522748

Fields of interest
Eggs, fish and fish products, meat and meat products, poultry, sensory analysis, nutrition, quality, preservation

KALAC, Pavel Prof., Ing., C.Sc.
University of South Bohemia
Department of Chemistry
Professor
Studenska 13
CZ - 37005 Ceske Budejovice
Tel: +42 38 5300404
Fax: +42 38 5300405
E-Mail: kalac@zf.jcu.cz

Private address
Cecova 56
CZ - 37004 Ceske Budejovice
Tel: +42 38 7330035

Fields of interest
Biogenic Amines, plant toxins, toxic trace elements, CE, GC, food composition

KAS, Jan Ph.D., D.Sc., Prof.
Institute of Chemical Technology
Department of Biochemistry and Microbiology
Head of Department
Technická 3
CZ - 16628 Praha 6
Tel: +420 2 24353018, 24353026
Fax: +42 2 3113726

E-Mail: jan.kas@vscht.cz

Private address
Koulova 6/1594
CZ - 16000 Praha 6
Tel: +42 2 3111965

Fields of interest
Proteins, enzymology, immunology, metabolism, biotechnology, irradiation

KELLNER, Vladimir M.Sc., Ph.D.
Research Institute of Brewing and Malting
Head of Analytical Division
Lipova 15
CZ - 12044 Praha 2
Tel: +420 2 295037, 293555
Fax: +420 2 291756

Private address
Makovskeho 1177/1
CZ - 16300 Praha 6

Fields of interest
Beer, additives, trace elements, biogenic amines, PAHs, PCBs, toxic trace elements, analytical chemistry, management

KUCERA, Jiri Ing.
Food Reserach Institute Prague
Head of Department of Microbial Products
Radiová 7
CZ - 10231 Praha 10
Tel: +42 2 702331 364
Fax: +42 2 701983
E-Mail: j.kucera@vupp.cz

Private address
Na pískách 95
CZ - 16600 Praha 6
Tel: +42 2 3113969

Fields of interest
Fish and fish products, meat and meat products, poultry, transgenic foods, additives, biopolymers, proteins, trace elements, biogenic amines, plant toxins, electrophoresis, HPLC, allergology, enzymology, food composition, biotechnology, fermentation, hygiene

KVASNICKA, Frantisek Ph.D.
ICT
Teacher, Researcher
Technicka 3
CZ - 16628 Praha
Tel: +420 2 24355117
Fax: +420 2 3116284
E-Mail: kvasnicf@vscht.cz

Private address
Silkova 550/6
CZ - 16900 Praha
Tel: +42 2 20516200
Fax: +420 2 20516200

Fields of interest
Fruits and vegetables, meat and meat products, poultry, additives, antioxidants, polyphenols, electrophoresis, HPLC, adulteration, analytical chemistry, preservation

MASKOVÁ, Eva Ing.
Food Research Institute Prague
Research Worker
Radiová 7
CZ - 10231 Praha 10
Tel: +420 2 702331 375, 283
Fax: +420 2 701983
E-Mail: e.maskova@vupp.cz

Private address
Bítovská 1226/7
CZ - 14000 Praha 4
Tel: +42 2 421185

Fields of interest
Vitamins, food composition, nutrition, high pressure technology

MELZOCH, Karel Dr.
Institute of Chemical Technology
Department of Fermentation Chemistry and Bioengineering
Associate Professor
Technická 5
CZ - 16628 Praha 6
Tel: +42 2 2435 4127
Fax: +42 2 24355051, 24311082, 3119990
E-Mail: karel.melzoch@vscht.cz

Private address
Borovanského 2379
CZ - 15500 Praha 5
Tel: +42 2 5619627

Fields of interest
Beverages, spirits, transgenic foods, wine, antioxidants, aroma active compounds, ethanol, polyphenols, mycotoxins, PAHs, electrophoresis, GC, HPLC, PCR, food composition, microbiology, nutrition, quality, biotechnology, fermentation, filtration, fouling and cleaning, HACCP

MIKOVA, Kamila Dipl.-Ing., C.Sc., Doc.
Boneco A.S.
Research and Development Manager, University Teacher
Kreslická 93
CZ - 10000 Praha 10

Tel: +42 2 67310499
Fax: +42 2 67311483

Private address
Evropská 26
CZ - 16000 Praha 6
Tel: +42 2 3115393

Fields of interest
Eggs, meat and meat products, poultry, GC, HPLC, food composition, microbiology, nutrition, quality, preservation, quality assurance, HACCP

PANOVSKÁ, Zdenka Dr. Ing.
Institute of Chemical Technology, Department of Food Chemistry and Analysis
Assistant
Technická 5
CZ - 16628 Praha 6
Tel: +420 2 24353263
Fax: +420 2 3119990
E-Mail: panovskz@vscht.cz

Private address
Muchova 15
CZ - 16000 Praha 6
Tel: +42 2 3120485

Fields of interest
Coffee, tea, cocoa, cosmetics, potatoes, sweeteners, sensory analysis, consumer research, food composition, rheology

PAVELKA, Jiri Dilp.-Ing., C.Sc.
Ekocentrum
Director
Martinovska ul.
CZ - 72308 Ostrava-Martinov
Tel: +42 69 6914646
Fax: +42 69 6914646
E-Mail: ekocentrum@ostrava.czcom.cz

Private address
Martinovská ul 261
CZ - 70800 Ostrava Plesná
Tel: +42 69 6957210

Fields of interest
Dairy, fish and fish products, meat and meat products, poultry, additives, biogenic amines, toxic trace elements, AAS, HPLC, analytical chemistry, food composition

PIPEK, Petr Ing., C.Sc.
Institute of Chemical Technology
Department of Food Preservation and Feed Technology
Associate Professor

Technická 3
CZ - 16628 Praha 6
Tel: +42 2 2435 3198
Fax: +42 2 3116284
E-Mail: petr.pipek@vscht.cz

Private address
Kurandové 9/672
CZ - 15200 Praha 5
Tel: +42 2 5819522
Fax: +42 2 5819522

Fields of interest
Fish and fish products, meat and meat products, poultry, additives, colours, proteins, biogenic amines, PAHs, management, microbiology, quality, drying, fermentation, hygiene, packaging, MAP, preservation, quality assurance, HACCP, storage, thermal processing

PLOCKOVÁ, Milada Assoc. Prof., Ph.D.
The Prague Institute of Chemical Technology
Assoc. Prof.
Technická 3
CZ - 16628 Praha
Tel: +420 2 24353275
Fax: +420 2 3119990
E-Mail: milada.plockova@vscht.cz

Private address
Na Cimelne 415/32
CZ - 15900 Prague

Fields of interest
Cosmetics, dairy, oils and fats, antimicrobials, biogenic amines, mycotoxins, electrophoresis, HPLC, PCR, microbiology, quality, fermentation, fouling and cleaning, hygiene, preservation, quality assurance

POKORNÝ, Jan Prof., Dipl.-Ing., Ph.D., D.Sc.
Prague Institute of Chemical Technology
Department of Food Chemistry and Analysis
Professor
Technická 5
CZ - 16628 Praha 6
Tel: +42 2 24353264
Fax: +42 2 3119990
E-Mail: jan.pokorny@vscht.cz

Private address
Kapradová 10
CZ - 10600 Praha 10
Tel: +42 2 72653919

Fields of interest
Oils and fats, antioxidants, aroma active compounds, lipids and fatty acids, oxidation reactions, polyphenols, HPLC, sensory analysis, nutrition, thermal processing

POUSTKA, Jan M.Sc., Ph.D.
Institute of Chemical Technology
Assistant Professor
Technická 3
CZ - 16628 Praha
Tel: +420 2 24353218
Fax: +420 2 24353185
E-Mail: jan.poustka@vscht.cz

Private address
Víta Nejedlého 14
CZ - 13000 Prague
Tel: +42 2 22722310

Fields of interest
Bread and cereals, fruits and vegetables, mycotoxins, PAHs, PCBs, pesticides, GC, HPLC, MS, analytical chemistry

PRUGAR, Jaroslav Dipl.-Ing., Dr.habil., Docent
Research Institute of Crop Production - Division of Plant Nutrition
Head of the Commission of Plant Products Quality
CAZV
Drnovska 507
CZ - 16106 Praha 6
Tel: +42 02 3302215
Fax: +42 02 33310636

Private address
Pocernicka 44
CZ - 10800 Praha 10
Tel: +42 2 74776937

Fields of interest
Bread and cereals, fruits and vegetables, potatoes, consumer research, food composition, nutrition, quality, quality of organic foods

PUDIL, Frantisek Dr.
Prague Institute of Chemical Technology
Department of Food Chemistry and Analysis
Assistant Professor

Technická 5
CZ - 16628 Praha 6
Tel: +42 2 24353183
Fax: +42 2 3119990
E-Mail: pudilf@vscht.cz

Private address
Renoirova 621/12
CZ - 15200 Praha 5
Tel: +42 2 5817965

Fields of interest
Aroma active compounds, chemometrics, GC, HPLC, MS, sensory analysis, food composition, quality, image processing information systems

RAUCH, Pavel Prof., Dipl.-Ing., Dr., D.Sc.
Institute of Chemical Technology
Department of Biochemistry and Microbiology
Head of Research Group
Technická 5
CZ - 16628 Praha 6
Tel: +420 2 24353076
Fax: +420 2 3119990
E-Mail: pavle.rauch@vscht.cz

Private address
Lumirova 5
CZ - 16628 Praha 2
Tel: +42 2 6927733

Fields of interest
Biopolymers, mycotoxins, pesticides, Immunoassays, analytical chemistry, bioavailability, enzymology, immunology, metabolism

RIHAKOVA, Zdenka M.Sc.
Institute of Chemical Technology
Assistent
Technickà 3
CZ - 16628 Praha
Tel: +420 2 24353822
Fax: +420 2 24353285
E-Mail: zdenka.rihakova@vscht.cz

Private address
Chemickà 953
CZ - 14000 Prague

Fields of interest
Cosmetics, oils and fats, antimicrobials, lipids and fatty acids, analytical chemistry, food composition, microbiology, biotechnology, preservation

RYCHTERA, Mojmír Prof., Ing., C.Sc.
Institute of Chemical Technology
Department of Fermentation Chemistry and Bioengineering
Professor
Technická 5
CZ - 16628 Praha 6
Tel: +42 02 3111509, 24354124
Fax: +42 02 24355051, 24311082
E-Mail: mojmir.rychtera@vscht.cz

Private address
Malenická 11/1790
CZ - 14800 Praha 4
Tel: +42 2 71913118

Fields of interest
Spirits, sugar, honey, sugar alcohols, wine, antimicrobials, carotenoids, ethanol, lactic acid, steroids, microbiology, biotechnology, fermentation

SCHWARZ, Walter Ing., C.Sc.
Setuza a.s.
Quality Manager
Zukova 100
CZ - 40129 Ústi nad Labem
Tel: +42 47 5292230
Fax: +42 47 5293924
E-Mail: walter.schwarz@setuza.cz

Private address
Visnová 2936/5
CZ - 40011Ústi nad Labem
Tel: +42 47 46568

Fields of interest
Oils and fats, antioxidants, fat replacers, lipids and fatty acids, management, quality, quality assurance, HACCP

STETINA, Jiri Ph.D.
The Prague Institute of Chemical Technology
Assistant Professor
Technická 3
CZ - 16628 Praha
Tel: +420 2 24353271
Fax: +420 2 3119990
E-Mail: jiri.stetina@vscht.cz

Private address
Spanielova 1290
CZ - 16300 Prague
Tel: +42 2 3019761

Fields of interest
Dairy, proteins, CE, electrophoresis, HPLC, sensory analysis, quality, rheology

SUIRAKOVÁ, Eva Ph.D.
The Prague Institute of Chemical Technology
Assistant
Technická 3
CZ - 16628 Praha
Tel: +420 2 24353271
Fax: +420 2 3119990
E-Mail: eva.suirakova@vscht.cz

Private address
Krupská 30
CZ - 10000 Prague
Tel: +42 2 7814172

Fields of interest
Dairy, antimicrobials, electrophoresis, PCR, microbiology, fermentation, freezing, hygiene, preservation

VALENTOVÁ, Helena Ing., C.Sc.
Institute of Chemical Technology, Department of Food Chemistry and Analysis

Assistant
Technická 5
CZ - 16628 Praha 6
Tel: +420 2 24353176
Fax: +420 2 3119990
E-Mail: valentoh@vscht.cz

Private address
Svatovitská 16
CZ - 16000 Praha 6
Tel: +42 2 24324694

Fields of interest
Beverages, cosmetics, meat and meat products, poultry, oils and fats, wine, sensory analysis, consumer research, food composition, rheology

---

VELISEK, Jan Prof., Ph.D., D.Sc.
Institute of Chemical Technology, Department of Food Chemistry and Analysis
Technická 1905
CZ - 16628 Praha 6
Tel: +42 02 24353177
Fax: +42 02 23119990
E-Mail: jan.velisek@vscht.cz

Private address
Kri zkovskeho 15
CZ - Praha 3
Tel: +42 2 6275192

Fields of interest
Amino acids, aroma active compounds, Maillard reaction products, vitamins, GC, HPLC, MS, sensory analysis, chemical reactions, food composition, toxicology

---

VLATER, Vladimír Dr.techn.,Dr.Sc.,Ing,Docent
Scientific Co-worker
Komroanská 30
CZ - 14319 Praha 4
Tel: +42 2 4024734
Fax: +42 2 4024030

Private address
5. Kvetna 25
CZ - 14000 Praha 4
Tel: +42 2 4397558

Fields of interest
Sugar, honey, sugar alcohols, Maillard reaction products, crystallisation

---

VODRAZKA, Zdenek Prof., Dipl.-Ing., Dr., Dr.Sc.
Department of Biochemistry and Microbiology
Technická 5
CZ - 16628 Praha 6
Tel: +42 02 24353357

Fax: +42 02 3119990

Private address
U smaltovny 19
CZ - 170 00 Praha
Tel: +42 2 800064

Fields of interest
Amino acids, biopolymers, enzymology, immunology, biotechnology, genetic engineering

Denmark

BALLING-ENGELSEN, Soren Ph.D.
The Royal Veterinary and Agricultural University
Department of Dairy and Food Science
Associate Professor
Rolighedsvej 30
DK - 1958 Frederiksberg C
Tel: +45 35 283205
Fax: +45 35 283245

Private address
Haraldslundvej 27
DK - 2800 Lyngby
Tel: +45 45 879747

Fields of interest
Bread and cereals, fish and fish products, meat and meat products, poultry, oils and fats, starch, water, carbohydrates, lipids and fatty acids, chemometrics, spectroscopy, food composition, molecular modelling

---

BJERGEGAARD, Charlotte Ph.D.
Royal Veterinary and Agricultural University
Chemistry Department
Research Scientist
Thorvaldsensvej 40
DK - 1871 Frederiksberg C
Tel: +45 35 282454
Fax: +45 35 282398
E-Mail: cbj@kvl.dk

Private address
Ejbyvej 60B
DK - 4623 Lille Skensved
Tel: +45 56 822484

Fields of interest
Antioxidants, carbohydrates, dietary fibres, CE, analytical chemistry, food composition, quality, natural product chemistry

---

CHRISTOPHERSEN, Carsten cand. scient.
University of Copenhagen
Lecturer

Universitetsparken 5
DK - 2100 Copenhagen
Tel: +45 35 320157
Fax: +45 35 320212
E-Mail: carsten@kiku.dk

Private address
Duevej 16, 2tv
DK - 2000 Copenhagen
Tel: +45 38 193260

Fields of interest
Fish and fish products, fruits and vegetables, antimicrobials, antioxidants, aroma active compounds, colours, lipids and fatty acids, oxidation reactions, polyphenols

HOLCH, Klaus Ph.D., M.Sc.
National Food Agency of Denmark
Institute of Food Chemistry and Nutrition
Head of Institute
Morkhoj Bygade 19
DK - 2860 Soborg
Tel: +45 33 956000
Fax: +45 33 956001
E-Mail: kh@fdir.dk

Fields of interest
Management

HOLMER, Gunhild Professor
Technical University of Denmark
Department of Biochemistry and Nutrition
Professor
Building 224
DK - 2800 Lyngby
Tel: +45 45 252736, 252731
Fax: +45 45 886307
E-Mail: hoelmer@mimer.be.dtu.dk

Private address
Kulsviertoften 43
DK - 2800 Lyngby
Tel: +45 45 882010
Fax: +45 45 882010

Fields of interest
Dairy, fish and fish products, oils and fats, lipids and fatty acids, oxidation reactions, vitamins, GC, HPLC, MS, nutrition

MELCHIOR-LARSEN, Lone Ph.D.
Royal Veterinary and Agricultural University
Department of Chemistry
Assoc. Professor
Thorvaldsensvej 40
DK - 1871 Frederiksberg C.
Tel: +45 35 282436
Fax: +45 35 282089

E-Mail: lone@kvl.kd

Fields of interest
Bread and cereals, fruits and vegetables, meat and meat products, poultry, potatoes, aroma active compounds, oxidation reactions, polyphenols, spectroscopy, analytical chemistry, enzymology

MUNCK, Lars Dr.Sci.Fil.Dr.
The Royal Veterinary and Agricultural University
Department of Dairy and Food Science
Professor, Head of Food Technology
Rolighedsvej 30
DK - 1958 Frederiksberg C.
Tel: +45 35 283358
Fax: +45 35 283245
E-Mail: lmu@kul.dk

Private address
Sostrade 5
DK - 3000 Helsingor
Tel: +45 49 200412

Fields of interest
Beer, bread and cereals, dairy, fish and fish products, meat and meat products, poultry, starch, sugar, honey, sugar alcohols, chemometrics, HPLC, spectroscopy, process monitor and quality control

OGAARD-MADSEN, Jorgen Ph.D.
The Technical University of Denmark
Department of Organic Chemistry
Senior Lecturer
Building 201
DK - 2800 Lyngby
Tel: +45 45 252157, 252155
Fax: +45 45 933968
E-Mail: okjom@pop.dtu.dk

Fields of interest
Spices, GC, MS, analytical chemistry

OTTE, Jeanette M.Sc., Ph.D.
The Royal Veterinary and Agricultural University
Department of Dairy and Food Science
Associate Professor
Rolighedsvej 30
DK - 1958 Frederiksberg C
Tel: +45 35 283189
Fax: +45 35 283190

Private address
Lyngbygardsvej 21A
DK - 2800 Lyngby
Tel: +45 21 771494
Fields of interest
Dairy, fat replacers, proteins, CE, HPLC, analytical chemistry, enzymology, food composition, quality,

biotechnology

ROSS-PETERSEN, Karl Jakob Ph.D.
KJ Ross-Petersen AS
Director
Agern Allé 3
DK - 2970 Horsholm
Tel: +45 45 865246
Fax: +45 45 865003
E-Mail: karl@ross.dk

Private address
Rydholt
DK - 2840 Holte
Tel: +45 45 425131

Fields of interest
Antioxidants

SKIBSTED, Leif H. Ph.D.
The Royal Veterinary and Agricultural University
Department of Dairy and Food Science
Professor of Food Chemistry
Rolighedsvej 30
DK - 1958 Frederiksberg C.
Tel: +45 35 283221
Fax: +45 35 283344

Private address
Orstedvej 54 B
DK - 4310 Viby

Fields of interest
Beverages, dairy, meat and meat products, poultry, antioxidants, carotenoids, lipids and fatty acids, oxidation reactions, quality, high pressure technology

---

SORENSEN, Susanne Ph.D.
Royal Veterinary and Agricultural University
Chemistry Department
Associate Professor
40, Thorvalsensvej
DK - 1871 Frederiksberg C
Tel: +45 35 282432
Fax: +45 35 282398
E-Mail: sus@kvl.dk

Private address
Oster Alle 2
DK - 2660 Brondby Strand
Tel: +45 43 734557

Fields of interest
Legumes, amino acids, antioxidants, proteins, CE, electrophoresis, adulteration, analytical chemistry, enzymology, food composition

---

SORENSEN, Hilmer Ph.D.
Royal Veterinary and Agricultural University
Chemistry Department
Associate Professor
40, Thorvalsensvej
DK - 1871 Frederiksberg C
Tel: +45 35 282432
Fax: +45 35 282398
E-Mail: hils@kvl.dk

Private address
Markvej 80
DK - 2660 Brondby Strand
Tel: +45 43 731490

Fields of interest
Amino acids, antioxidants, carbohydrates, dietary fibres, proteins, CE, adulteration, analytical chemistry, chemical reactions, enzymology, food composition, metabolism, natural products, chemistry and biorefining in pilot plant scale

Estland

ILMOJA, Kalle
Health Protection Inspectorate, Central Laboratory
Heat of Laboratory Chemistry Tartu
Pöllu 1a
EE - 51009 Tartu
Tel: +372 7 447422
Fax: +372 7 447422
E-Mail: ilmoja@ttkt.tartu.ee

Private address
Kalevi 71-12
EE - 51008 Tartu
Tel: +372 7 471793

Fields of interest
Beverages, fish and fish products, fruits and vegetables, meat and meat products, poultry, potatoes, water, additives, antimicrobials, antioxidants, colours, minerals, trace elements, alkaloids, biogenic amines, mycotoxins, PAHs, PCBs, pesticides, plant toxins, toxic trace elements, AAS, GC, HPLC, MS, polarography, sensory analysis, TLC, adulteration, analytical chemistry, consumer research, food composition, quality, toxicology, quality assurance, HACCP

VOKK, Raivo Ph.D.
Tallin Technical University
Department of Food Processing
Head of the Department, Chairman
Ehitajate tee 5
EE - 19086 Tallinn
Tel: +37 2 6202952
Fax: +37 2 6202950

Private address
Raja 7a II

EE - 11219 Tallinn
Tel: +37 2 25153245

Fields of interest
Antimicrobials, antioxidants, aroma active compounds, dietary fibres, sensory analysis, nutrition, hygiene, rheology

Finland

AHOLA, Maarit M.Sc.
Finpro
Senior Consultant
P.O.Box 908
FIN - 00101 Helsinki
Tel: +358 204 695225
Fax: +358 204 695456
E-Mail: maarit.ahola@finpro.fi

Private address
Tyynelänkuja 5 B 18
FIN - 00780 Helsinki
Tel: +358 0 3451267

Fields of interest
Consumer research, management, nutrition

AHVENAINEN, Juha M.I. Ph.D.
VTT, Food Research Laboratory
VTT Biotechnology
Institute Director
Tietotie 2, P.O.Box 1500
FIN - 02044 VTT, Espoo
Tel: +358 0 4565160
Fax: +358 0 4552103
E-Mail: juha.ahvenainen.vtt.fi

Private address
Raiviontie 6
FIN - 00670 Helsinki
Tel: +358 0 7243263
Fax: +358 0 7243263

Fields of interest
Beer, bread and cereals, starch, management, biotechnology

ANDERSSEN, Valborg M.Sc.
Valio Ltd.
Dairy Industry
Meijeritie 3
FIN - 00039 Valio (Helsinki)
Tel: +358 20 381121
Fax: +358 10 3812737
E-Mail: valborg.anderssen @valio.fi

Fields of interest
Beverages, dairy, management, microbiology, quality, hygiene, quality assurance, HACCP

APPLEBYE, Ulla Ph.D.
Valio Ltd.
Dairy
P.O.Box 30
FIN - 00039 Valio
Tel: +358 10 3813314
Fax: +358 10 3813049
E-Mail: ulla.applebye@valio.fi

Fields of interest
Dairy, sensory analysis, consumer research, quality

ARO, Tarja M.Sc.
University of Turku
Department of Biochemistry and Food Chemistry
Research Chemist
FIN - 20014 Turku
Tel: +358 2 3336876
Fax: +358 2 3336860
E-Mail: tarja.aro@utu.fi

Fields of interest
Fish and fish products, aroma active compounds, lipids and fatty acids, GC

AURA, Anna-Marija Licentiate of Technology
VTT Biotechnology
Research Scientist
Tietotie 2, P.O.Box 1500
FIN - 02044 VTT, Espoo
Tel: +358 9 4566178
Fax: +358 9 4552103
E-Mail: anna-marija.aur@vtt.fi

Private address
Hyljelahdentie 19 as 1
FIN - 02260 Espoo
Tel: +358 9 8043585

Fields of interest
Bread and cereals, Coffe, tea, cocoa, fruits and vegetables, berries, starch, wine, antioxidants, carbohydrates, carotenoids, dietary fibres, polyphenols, bioavailability, food composition, metabolism, nutrition, high pressure technology, minimal processing, packaging, MAP, thermal processing

EERIKÄINEN, Tero Dr.techn. (Chem. Eng.)
Helsinki University of Technology
Laboratory of Bioprocess Engineering
Senior Assistant
Kemistintie 1
FIN - 02015 HUT

Tel: +358 451 2559
Fax: +358 0 462373
E-Mail: tero.eerikainen@hut.fi

Private address
Kirjailijantie 5
FIN - 05830 Hyvinkää

Fields of interest
Sugar, honey, sugar alcohols, biopolymers, chemometrics, enzymology, metabolism, quality, biotechnology, fermentation

HÄGG, Margareta Ph.D., B.L.
Centre for Metrology and Accreditation
Lead Assessor
FIN - 00181 Helsinki
Tel: +358 9 6167236
Fax: +358 9 6167341
E-Mail: margareta.hagg@mikes.fi

Private address
Östra Oxgränden 9
FIN - 02580 Sjundea
Tel: +358 40 5514272

Fields of interest
Fruits and vegetables, water, antioxidants, carotenoids, vitamins, HPLC, food composition, quality, quality assurance, HACCP

HAKALA, Mari M.Sec.
University of Turku
Department of Biochemistry and Food Chemistry
Research Chemist
FIN - 20014 Turku
Tel: +358 2 3336874
Fax: +358 2 3336860
E-Mail: mari.hakala@utu.fi

Fields of interest
Fruits and vegetables, aroma active compounds, GC, MS, sensory analysis, analytical chemistry, quality

HÄKKINEN, Sari M.Sc.
University of Kuopio
Department of Applied Chemistry and Microbiology
Research Chemist
P.O.Box 27
FIN - 00014 University of Helsinki
Tel: +358 9 19958234
Fax: +358 9 19158475
E-Mail: sari.h.hakkinen@helsinki.fi

Fields of interest
Beverages, fruits and vegetables, wine, antioxidants, carotenoids, polyphenols, AAS, GC, HPLC, MS, sensory analysis, spectroscopy, food composition, nutrition, freezing, storage, thermal processing

HEINIÖ, Raija Liisa M.Sc.
VTT Biotechnolgoy
Senior Research Scientist
Tietotie 2, P.O.Box 1500
FIN - 02044 VTT, Espoo
Tel: +358 9 4565178
Fax: 358 9 4552103
E-Mail: raija-liisa.heinio@vtt.fi

Private address
Pihkapolku 1 F
FIN - 02110 Espoo

Fields of interest
Bread and cereals, aroma active compounds, sensory analysis, sensor technology, consumer research, quality

HEINONEN, Marina Ph.D.
University of Helsinki
Department of Applied Chemistry and Microbiology
Assistant Professor
P.O. Box 27
FIN - 00014 Helsinki
Tel: +358 9 19158446
Fax: +358 9 19158475
E-Mail: marina.heinonen@helsinki.fi

Private address
Klovirinne 5
FIN - 02180 Espoo
Tel: +358 9 5022814

Fields of interest
Dairy, fruits and vegetables, meat and meat products, poultry, oils and fats, potatoes, antioxidants, carotenoids, colours, lipids and fatty acids, polyphenols, vitamins, HPLC, analytical chemistry, food composition

HIETANIEMI, Veli M.Sc.
Agricultural Research Centre of Finland, Food Research, Chemistry Laboratory
Laboratory Manager
FIN - 31600 Jokioinen
Tel: +358 3 41881
Fax: +358 3 41883266
E-Mail: veli.hietaniemi@mtt.fi

Private address
Otsontie 6
FIN - 31600 Jokioinen
Tel: +358 3 4384416

Fields of interest
Bread and cereals, antioxidants, mycotoxins, PAHs, PCBs, GC, HPLC, MS, food composition, high pressure technology

HOME, Silja D. Tech. (Chem. Eng.)
VTT Biotechnology and Food Research
Senior Scientist
P.O. Box 1500
FIN - 02044 VTT (Espoo)
Tel: +358 9 4565115
Fax: +358 9 4552028
E-Mail: silja.home@vtt.fi

Private address
Lystimäenkuja 5
FIN - 02210 Espoo
Tel: +358 9 889260
Fax: +358 9 8036004

Fields of interest
Beer, starch, carbohydrates, analytical chemistry, quality, biotechnology, fermentation

---

HONKAVAARA, Markku Ph.D.
Finnish Meat Research Institute
Research den Development
P.O.Box 56
FIN - 13101 Hämeentinna
Tel: +358 3 5705342
Fax: +358 3 5705329
E-Mail: markku.honkavaara@lth.htk.fi

Fields of interest
Meat and meat products, antioxidants, lipids and fatty acids, vitamins, nutrition, meat processing

---

HOPIA, Anu M.Sc., Ph.D.
University of Helsinki
Department of Applied Chemistry and Microbiology
Research Scientist
P.O. Box 27
FIN - 00014 Helsinki University
Tel: +358 9 19158249
Fax: +358 9 19158475
E-Mail: anu.hopia@helsinki.fi

Fields of interest
Fruits and vegetables, oils and fats, additives, antioxidants, aroma active compounds, colours, lipids and fatty acids, oxidation reactions, polyphenols, HPLC

HORNE-EKMAN, Maarit M.SC.
Pyhäjärvi Institute
Researcher
Ruukinpuisto
FIN - 27500 Kauttua
Tel: +358 2 8380615
Fax: +358 2 8380660
E-Mail: maarit.horne-ekman@pyhajarvi-inst.fi

Fields of interest
Fruits and vegetables, nutrition, drying, fermentation

HUOPALAHTI, Rainer Ph.D.
University of Turku
Department of Biochemistry and Food Chemistry
Senior Assistant
FIN - 20014 University of Turku
Tel: +358 21 6336872
Fax: +358 21 6336860
E-Mail: rainer.huopalahti@utu.fi

Private address
Kvartsikuja 3
FIN - 20740 Turku
Tel: +358 21 2423156
Fax: +358 21 2423155

Fields of interest
Eggs, fruits and vegetables, potatoes, spices, aroma active compounds, colours, hormones, mycotoxins, PAHs, plant toxins, CE, GC, HPLC, MS, analytical chemistry, food composition, SFE

JÄRVENPÄÄ, Eila M.Sc., Ph.D.
University of Turku
Department of Biochemistry and Food Chemistry
Research Chemist
FIN - 20014 Turku
Tel: +358 2 3336840
Fax: +358 2 3336860
E-Mail: eila.jarvenpaa@utu.fi

Private address
Annakuja 2
FIN - 21420 Lieto

Fields of interest
Eggs, fruits and vegetables, GC, HPLC, analytical chemistry, food composition, high pressure technology

---

JÄRVI-KÄÄRIÄINEN, Irma Terhen M.Sc.
Association of Packaging Technology and Research
Research Scientist
Mannerheimintie 156
FIN - 00270 Helsinki
Tel: +358 0 643497
Fax: +358 0 643498
E-Mail: ptr.ry@kolumbus.fi

Fields of interest
Consumer research, regulative issues, packaging, MAP, storage, food packaging, packaging & environment, active packaging, intelligent packaging

---

KALLIO, Heikki Ph.D.
University of Turku
Department of Biochemistry and Food Chemistry
Professor of Food Chemistry
FIN - 20014 University of Turku
Tel: +358 21 6336870

Fax: +358 21 6336860
E-Mail: heikki.kallio@utu.fi

Private address
Tammikuja 3
FIN - 21110 Naantali
Fax: +358 2 4353227

Fields of interest
Fruits and vegetables, oils and fats, aroma active compounds, lipids and fatty acids, GC, HPLC, MS, food composition, nutrition, supercritical fluid technologies

KEURULAINEN, Ritva M.Sc.
Valio Oy
Research and Devlopment Manager
P.O.Box 30
FIN - 73101 Lapinlahti
Tel: +358 10 3815373
Fax: +358 10 3815364
E-Mail: ritva.keurulainen@valio.fi

Fields of interest
Dairy, microbiology, nutrition, quality, regulative issues, drying, hygiene

KIVISTÖ, Laura M.Sc.
Atria Oyj
Microbiologist
Pl 117, Lapuantie 594
FIN - 60101 Seinäjoki
Tel: +358 6 4168111
Fax: +358 6 4168511
E-Mail: laura.kivisto@atria.fi

Private address
Niemistöntie 21
FIN - 60420 Seinäjoki
Tel: +358 6 4170679

Fields of interest
Meat and meat products, food composition, microbiology, quality, hygiene, quality assurance, HACCP

KORHONEN, Hann Ph.D.
Agricultural Research Centre of Finland, Food Research
Director, Professor
FIN - 31600 Jokioinen
Tel: +358 3 41883271
Fax: +358 3 41883244
E-Mail: hannu.korhonen@mtt.fi

Private address
28 A 3
FIN - 11120 Riihimähi
Tel: +358 19 732262

Fields of interest
Dairy, antimicrobials, antioxidants, polyphenols, proteins, immunology

KOSKENKORVA, Anneli M.Sc.
Food & Quality
Managing Director
Souttisalontie 11
FIN - 60510 Hyllykallio
Tel: +358 6 4128620
Fax: +358 6 4128621
E-Mail: anneli.koskenkorva@nic.fi

Fields of interest
Beverages, bread and cereals, cosmetics, fruits and vegetables, additives, microbiology, nutrition, biotechnology, minimal processing, quality assurance, HACCP

KUUSISTO, Päivi M.Sc.
Raisiogroup
Research Chemist
Raisionkaari 55, P.O.Box 101
FIN - 21201 Raisio
Tel: +358 2 4432558
Fax: +358 2 4432092
E-Mail: paivi.kuusisto@raisiogroup.com

Fields of interest
Oils and fats, lipids and fatty acids, GC, HPLC, analytical chemistry, quality, quality assurance, HACCP

LAAKSO, Päivi Ph.D.
Raisio Benecol Ltd
Research Chemist
Raisionkaari 55, P.O.Box 101
FIN - 21201 Raisio
Tel: +358 2 4432404
Fax: +358 2 4432092
E-Mail: raivi.laakso@raisiogroup.com

Fields of interest
Oils and fats, lipids and fatty acids, GC, HPLC, MS, TLC, analytical chemistry

LAMPI, Anna-Maija D.Sc, M.Sc.
University of Helsinki
Researcher
P.O. Box 27, Viikki-D
FIN - 00014 Helsinki
Tel: +358 9 19158412
Fax: +358 9 19158475
E-Mail: anna-maija.lampi@helsinki.fi

Private address
Petaksentie 39
FIN - 00630 Helsinki

Tel: +358 9 7242402

Fields of interest
Oils and fats, lipids and fatty acids, oxidation reactions, steroids, vitamins, GC, HPLC

LAMPOLAHTI, Soili M.Sc.
National Veterinary And Food Research Institute (EELA)
P.O.Box 368, Hämeentie 57
FIN - 00231 Helsinki
Tel: +358 9 393101
Fax: 358 9 3931920
E-Mail: soili.lampolahti@eela.fi

Fields of interest
Dairy, fish and fish products, sensory analysis, quality

LAPVETELÄINEN, Anja M.Sc., D.Ph.
University of Kuopio
Centre for Training and Development
Sensory Specialist
P.O. Box 6127
FIN - 70211 Kuopio
Tel: +358 17 163931
Fax: +358 17 163903
E-Mail: anja.lapvetelainen@uku.fi

Fields of interest
Bread and cereals, fish and fish products, fruits and vegetables, proteins, sensory analysis

LEHTONEN, Pekka Ph.D.
Alko Group LTD., Alcohol Control Laboratory
Laboratory Director
P.O. Box 279
FIN - 01301 Vantaa, Helsinki
Tel: +358 9 57655389
Fax: +358 9 57655252
E-Mail: pekka.lehtonen@alko.fi

Fields of interest
Beer, beverages, spirits, wine, additives, carbohydrates, ethanol, polyphenols, biogenic amines, drugs, mycotoxins, AAS, GC, HPLC, MS, analytical chemistry, food composition, quality

MÄÄTTÄ, Kaisu M.Sc.
University of Kuopio
Research Chemist
P.O.Box 17´627
FIN - 70211 Kuopio
Tel: +358 17 163103
Fax: +358 17 163112
E-Mail: kaisu.maatta@uku.fi

Fields of interest
Fruits and vegetables, antioxidants, GC, HPLC, MS, analytical chemistry, food composition

MÄKINEN, Marjukka M.Sc. in Food Chemistry
University of Helsinki
Department of Applied Chemistry and Microbiology
Researcher
P.O. Box 27
FIN - 00014 University of Helsinki
Tel: +358 9 1911
Fax: +358 9 58475
E-Mail: marjukka.makinen@helsinki.fi

Fields of interest
Oils and fats, antioxidants, lipids and fatty acids, oxidation reactions, HPLC, chemical reactions

MATILAINEN, Katri M.Sc.(Chem.Eng.)
Valio Ltd.
R & D Cheese
Meijeritie 4, P.O.Box 30
FIN - 00039 Valio
Tel: +358 10 3812515
Fax: +358 10 3832515
E-Mail: katri.matilainen@valio.fi

Fields of interest
Beer, dairy, food composition, microbiology, quality, biotechnology

MOILANEN, Raija Ph.D.
Finnfeeds Finland Ltd.
Development
Satamatie 2
FIN - 21100 Naantali
Tel: +358 2 4393314
Fax: +358 2 4393333
E-Mail: raija.moilanen@danisco.com

Private address
Paimentie 6 A 4
FIN - 20760 Piispanristi

Fields of interest
Sugar, honey, sugar alcohols, carbohydrates, vitamins, HPLC, analytical chemistry, microbiology, quality, biotechnology, hygiene, quality assurance, HACCP

NIEMI, Sanna-Maria M.Sc., Dietition
Fazer Bakeries Ltd.
Nutritionist
P.O.Box 40
FIN - 15101 Lahti
Tel: +358 3 8571562
Fax: +358 3 8571371
E-Mail: sanna-maria.niemi@fazer.fi

Fields of interest
Bread and cereals, dietary fibres,

nutrition

OLLILAINEN, Velimatti M.Sc.
University of Helsinki
Department of Applied Chemistry and Microbiology
Assistant in Food Chemistry
Latokartanonkaari 11, P.O. Box 27
FIN - 00014 University of Helsinki
Tel: +358 9 1911
Fax: +3589 19158475
E-Mail: velimatti.ollilainen@helsinki.fi

Private address
Kuusmiehentie 28 B5
FIN - 00670 Helsinki
Tel: +358 9 7249816
Fax: +358 9 7249816

Fields of interest
Vitamins, HPLC, MS, quality assurance in laboratory, analytical chemistry, food composition, quality

PIIRONEN, Vieno Ph.D.
University of Helsinki
Department of Applied Chemistry and Microbiology
Professor
P.O. Box 27
FIN - 00014 Helsinki
Tel: +358 9 19158222
Fax: +358 9 19158475

Private address
Vihertie 64D
FIN - 01620 Vantaa
Tel: +358 9 890760

Fields of interest
Lipids and fatty acids, oxidation reactions, steroids, vitamins, HPLC, analytical chemistry, chemical reactions, food composition

RANTAMÄKKI, Pirjo Dr.
Agricultural Research Centre of Finland
Research Scientist
Myllytie 1
FIN - 31600 Jokioinen
Tel: +358 3 41881
Fax: +358 3 41883244
E-Mail: pirjo.rantamaki@mtt.fi

Fields of interest
Dairy, proteins, PAHs, thermal anlaysis, functional properties of proteins

RÄSÄNEN, Janne D.Sc.
Vaasan & Vaasan Oy
Development Manager
Nuijalantie 11
FIN - 02630 Espoo
Tel: +358 204 462337
Fax: +358 204 462340
E-Mail: janne.rasanen@vaasan.com

Fields of interest
Bread and cereals, starch, water, dietary fibres, thermal anlaysis, management, fermentation, freezing, rheology, storage

RICHTER, Timo M.Sc.
Oy Henkel-Ecolab Ab
Industry Manager
Mäkelänkatu 54 A
FIN - 00510
Tel: +358 9 3965520
Fax: +358 9 39655306

Fields of interest
Beer, water, management, microbiology, biotechnology, hygiene

SAARINEN, Niina FM
University of Turku
Institute of Biomedicine
Researcher
Kiinamyllynkatu 10
FIN - 20520 Turku
Tel: +358 2 33371
Fax: +358 2 3337352
E-Mail: nisaarin@utu.fi

Fields of interest
Fruits and vegetables, dietary fibres, polyphenols, steroids, hormones, bioavailability, metabolism

SALMINEN, Seppo M.Sc., (Food Science + Food Chemistry), Ph.D.
University of Turku
Department of Biochemistry and Food Chemistry
Professor, Food Development, Visiting Professor of Food Toxicology RMIT University, Melbourne, Australia
FIN - 20014 Turku
Tel: +358 400 601394
Fax: +358 2 3336860
E-Mail: seppo.salminen@utu.fi

Fields of interest
Additives, carbohydrates, mycotoxins, microbiology, nutrition, toxicology

SAUKKO, Maire M.Sc.
The Vocational School of Northern Ostrobothnia
Head of Processing Department
Isokatu 1, P.O.Box 325
FIN - 90101 Oulu
Tel: +358 8 3128311

Fax: +358 8 3128528
E-Mail: maire.saukko@osakk.fi

Private address
Alitie 6B45
FIN - 90100 Oulu
Tel: +358 8 330148

Fields of interest
Bread and cereals, fruits and vegetables, additives, biopolymers, sensory analysis, food composition, microbiology, hygiene, minimal processing, quality assurance, HACCP

SILANPÄÄ, Rauno M.Sc.
Oy Sinebrychoff Ab
Process Manager
P.O.Box 87
FIN - 04201 Kerava
Tel: +358 9 294991
Fax: +358 9 29499408
E-Mail: rauno.sillanpaa@koff.fi

Fields of interest
Beer, beverages

SKRÖKKI, Leila Anneli Phil. Lis.
Chemist

Private address
Yrsonranta IA6
FIN - 90240 Oulo
Tel: +358 50 3379756

Fields of interest
Meat and meat products, poultry, oils and fats, sugar, honey, sugar alcohols, water, adulteration, food composition, quality

STORGARDS, Erna Dr.
VTT Biotechnolgoy
Research Microbiologist
Tietotie 2, P.O.Box 1500
FIN - 02044 VTT, Espoo
Tel: +358 9 4564526
Fax: +358 9 4562103
E-Mail: erna.storgards@vtt.fi

Private address
Kiiltokallionkuja 8 B
FIN - 02180 Espoo
Tel: +358 9 4208410

Fields of interest
Beer, beverages, fouling and cleaning, hygiene, quality assurance, HACCP

SUUTARINEN, Marjaana M.Sc., Licentiate in Technology
VTT Biotechnology and Food Research
Research Scientist
P.O. Box 1500
FIN - 02044 Espoo
Tel: +358 0 4565197
Fax: +358 0 4552103
E-Mail: marjaana.suutarinen@vtt.fi

Fields of interest
Bread and cereals, fruits and vegetables, chemometrics, sensory analysis, spectroscopy, analytical chemistry, food composition, freezing, minimal processing

TAHVONEN, Raija Ph.D.
University of Turku
Department of Biochemistry and Food Chemistry
Vatselankatu 2
FIN - 20014 Turku
Tel: +358 2 3336844
Fax: +358 2 3336790
E-Mail: raija.tahvonen@utu.fi

Fields of interest
Lipids and fatty acids, minerals, AAS, nutrition

TIUSANEN, Sirkka M.Sc.
Porilab
Controlling of Food and Environment
Tiedepuisto 4
FIN - 28600 Pori
Tel: +358 26 213341
Fax: +358 26 213333
E-Mail: sirkka.tiusanen@pori.fi

Private address
Suosmeri
FIN - 28450 Ulvila
Tel: +358 25 386075

Fields of interest
Meat and meat products, poultry, water, additives, toxic trace elements, AAS, analytical chemistry

TOIKKANEN, Kari M.Sc. (Dairy Technology) eMBA
Valio Ltd.
Product Development Manager
Meijeritie 4, P.O.Box 30
FIN - 00039 Valio
Tel: +358 10 381121
Fax: +358 10 3813129
E-Mail: kari.toikkanen@valio.fi

Fields of interest
Dairy, fruits and vegetables, sensory analysis, management, packaging, MAP, functional foods

TOIVO, Jari M.Sc.

University of Helsinki
Researcher
P.O.Box 27, Viikki - D
FIN - 00014 Helsinki
Tel: +358 9 19158407
Fax: +358 9 19158475
E-Mail: jari.toivo@helsinki.fi

Private address
Saagatie 10 D 18
FIN - 01200 Vantaa
Tel: +358 9 8768376

Fields of interest
Lipids and fatty acids, steroids, GC, TLC, analytical chemistry, bioavailability, food composition

TUOMALA-SARAMÄKI, Terhi M.Sc.
Novolab Oy
Department Manager
Lepolantie 5
FIN - 03600 Karkkila
Tel: +358 9 2258610
Fax: +358 9 2259785
E-Mail: terhi.tuomala-saramaki@novolab.fi

Fields of interest
Microbiology, quality, hygiene, quality assurance HACCP

TYKKYLÄINEN, Paavo M.Sc.
Valio Ltd.
Product Development Manager
Meijeritie 4, P.O.Box 30
FIN - 00039 Valio
Tel: +358 10 3813338
Fax: +358 10 3813049
E-Mail: paavo.tykkylainen@valio.fi

Private address
Pormestarinrinne 2 B 19
FIN - 00160 Helsinki
Tel: +358 9 658031

Fields of interest
Dairy, dietary fibres, fat replacers, minerals, microbiology, hygiene, preservation, thermal processing, probiotics, functional dairy products

VAHTERISTO, Liisa Ph.D.
University of Helsinki
Department of Applied Chemistry and Microbiology
Research Scientist, Project Leader
Latokartanonkaari 11, P.O. Box 27
FIN - 00014 University of Helsinki
Tel: +358 9 19158400
Fax: +358 9 19158475
E-Mail: liisa.vahteristo@helsinki.fi

Fields of interest
Vitamins, HPLC, analytical chemistry, bioavailability, food composition, nutrition

France

BIRLOUEZ, Inès Dr., Ing.
Institut National Agromonique, Laboratoire de Chimie Analytique
Maitre de Conférence
16, Rue Claude-Bernard
F - 75005 Paris
Tel: +33 1 44081649
Fax: +33 1 44081653
E-Mail: birlouezie@aol.com

Private address
Rue Héléne Boucher
F - 95120 Ermont
Tel: +33 1 34145840
Fax: +33 1 34145056

Fields of interest
Dairy, fruits and vegetables, amino acids, antioxidants, Maillard reaction products, oxidation reactions, proteins, vitamins, HPLC, spectroscopy, analytical chemistry, nutrition, quality, minimal processing, thermal processing

CHEFTEL, Jean-Claude Dr., Prof.
Université des Sciences et Techniques Montpellier II
Unité de Biochimie-Technologie Alimentaires
Place Eugène-Bataillon
F - 34095 Montpellier Cedex 5
Tel: +33 67 143351
Fax: +33 67 633397
E-Mail: cheftel@arpb.univ-montp2.fr

Fields of interest
Proteins, high pressure technology, minimal processing, electric pulse processing

DECARIS, Bernard Prof.
Université de Nancy I
Laboratoire de Génétique (associé INRA)
Rue Lionnois
F - 54000 Nancy
Tel: +33 83 912096
Fax: +33 83 912500
E-Mail: decaris@nancy.inra.fr
Private address
7, Rue Lacretelle
F - 54000 Nancy
Tel: +33 83 951026

Fields of interest
Dairy, transgenic foods, antimicrobials,

carbohydrates, electrophoresis, PCR, microbiology, nutrition, genetic engineering

DUCAUZE, Christian J. Prof., Dr.
Institut National Agronomique Paris-Grignon
16, Rue Claude Bernard
F - 75231 Paris Cedex 5
Tel: +33 1 44081644
Fax: +33 1 44081653

Fields of interest
Water, wine, aroma active compounds, trace elements, toxic trace elements, chemometrics, GC, analytical chemistry, quality, quality assurance, HACCP

FEINBERG, Max Dr., Ph.D.
Institut National de la Recherche Agronomique,
Analytical Chemistry
Director
Rue Claude-Bernard, 16
F - 75231 Paris Cedex 5
Tel: +33 1 44081652
Fax: +33 1 44081653
E-Mail: feinberg@inapg.inra.fr

Private address
Bd. Kellermann, 100
F - 75013 Paris
Tel: +33 1 45651425

Fields of interest
Maillard reaction products, chemometrics, HPLC, sensory analysis, adulteration, analytical chemistry, quality, NMR

GAUCHERON, Frédéric Dr.
INRA
Research
65, Rue de Saint-Brieuc
F - 35042 Rennes
Tel: +33 99 285342
Fax: +33 99 285350
E-Mail: fgauchez@rennes.inra.fr

Fields of interest
Dairy, minerals, proteins, trace elements, AAS, HPLC, spectroscopy, adulteration, analytical chemistry, chemical reactions, food composition, quality, high pressure technology, storage, thermal processing

GRAILLE, Jean Ph.D., Ing.
CIRAD-CP, Laboratoire de Lipotechnie
CIRAD-AMIS
Head of the LIPOTECHNY LABORATORY
2477, Av. du Val de Montferrand, P.O. Box 5035
F - 34032 Montpellier Cedex 5
Tel: +33 467 615881
Fax: +33 467 615955
E-Mail: jean.graille@cirad.fr

Private address
Le Clos de la Belle 331 rue Amy Mollisson
F - 34070 Montpellier
Tel: +33 467 694101

Fields of interest
Oils and fats, additives, antioxidants, carotenoids, lipids and fatty acids, polyphenols, vitamins, PAHs, GC, HPLC, TLC, analytical chemistry, chemical reactions, biotechnology

ILARI, Jean-Luc Dr., Ing., H.D.R., Prof.
E.N.I.T.I.A.A. de Nantes
Ecole Nationale d' Ingenieurs des Techniques des I.A.A.
Prof. de Génie Alimentaire et Particulaire
La Géraudière, B.P. 82225
F - 44072 Nantes Cedex
Tel: +33 2 51785462
Fax: +33 2 51785455, 785467
E-Mail: ilari@enitiaa-nantes.fr

Private address
3, Allée des Mûriers
F - 44240 La Chapelle-Sur-Erdre
Tel: +33 2 40725181

Fields of interest
Additives, biopolymers, carbohydrates, fat replacers, proteins, chemometrics, drying, rheology, storage, powder technology and characterization

LE BOTLAN, Denis Dr.
C.N.R.S.
Faculty of Sciences
2, Rue de la Houssinière
F - 44072 Nantes Cedex
Tel: +33 251 125710
Fax: +33 251 125712
E-Mail: denis.lebotlan@chimbio.univ-nantes.fr

Private address
6, Rue Saint Saens
F - 44470 Carquefou
Tel: +33 240 509883

Fields of interest
Oils and fats, starch, water, carbohydrates, lipids and fatty acids, spectroscopy, analytical chemistry

LIDDLE, Peter M.A.
Bacardi - Martini Services
Group Scientific Coordinator (Europe)
P.O. Box 50
F - 93401 Saint - Ouen Cedex

Tel: +33 1 49454873
Fax: +33 1 49454905
E-Mail: peliddle@bacardi.com

Private address
2, Rue du Vivier
F - 95450 Ableiges

Fields of interest
Spices, spirits, wine, aroma active compounds, ethanol, GC, MS, adulteration, analytical chemistry, regulative issues

MARIN, Michèle Ph.D.
Institut National Agronomique Paris-Grignon
Professor of University
C.B.A.I.
F - 78850 Thiverval-Grignon
Tel: +33 1 30815439
Fax: +33 1 30815597
E-Mail: marin@platon.grignon.inra.fr

Private address
2, Rue des Marais
F - 78220 Viroflay
Tel: +33 1 30243467

Fields of interest
Aroma active compounds, ethanol, thermal anlaysis, drying, filtration, freezing, freeze-drying, preservation, separation, membranes, extraction

MARTIN, Gérard Dr.
Eurofins Scintific
Consultant
BP 42301
F - 44323 Nantes
Tel: +33 251 832113
Fax: +33 251 832110
E-Mail: gejemartin@aol.com

Private address
21, Rue Thomas-Maisonneuve
F - 44000 Nantes
Tel: +33 240 290066
Fax: +33 240 291799

Fields of interest
Fruits and vegetables, spirits, sugar, honey, sugar alcohols, wine, additives, chemometrics, MS, adulteration, analytical chemistry, NMR

MEUNIER, Jean-Claude Ph.D.
Institut National Agronomique Paris-Grignon
Professor
F - 78850 Thiveral - Grignon
Tel: +33 1 30815464
Fax: +33 1 30815373
E-Mail: meunier@platon.grignon.inra.fr

Private address
30, Rue E. Manet
F - 78370 Plaisir
Tel: +33 1 30559359

Fields of interest
HPLC, PCR, spectroscopy, enzymology, biotechnology

O´BRIEN, John B.Sc., M.Sc., Ph.D., EurTox.
Groupe Danone
Directeur, Securité des Aliments
15 Av. Galilée
F - 92350 Le Plessis Robinson
Tel: +33 1 41078801
Fax: +01 41 074775

Fields of interest
Dairy, heterocyclic aromatic amines, pesticides, plant toxins, chemical reactions, management, metabolism, nutrition, regulative issues, toxicology

RAHALI, Véronique Ing.
Groupe Ecole Superieure d' Agriculture
Teacher (Biochemistry) and Researcher
55, Rue Rabelais B.P. 748
F - 49007 Angers Cedex
Tel: +33 41 235555
Fax: +33 41 235500
E-Mail: v.rahali@esa-angers.educagri.fr

Private address
15, Rue du Marquis de Turbilly
F - 49000 Angers

Fields of interest
Dairy, proteins, electrophoresis, HPLC, spectroscopy, analytical chemistry, food composition, nutrition, quality, foaming and emulsiffying properties of proteins

ROLLIN, Patrick Dr., Prof.
LCBA, Université d'Orléans
ICOA
P.O. Box 6759
F - 45067 Orleans Cedex2
Tel: +33 238 417370
Fax: +33 338 494579

Private address
4, Rue Sous les Saints
F - 45000 Orleans
Tel: +33 038 680366

Fields of interest
Cosmetics, fruits and vegetables, legumes, antioxidants, aroma active compounds, carbohydrates, oxidation reactions, alkaloids, plant toxins, HPLC, MS, spectroscopy, TLC, analytical chemistry,

chemical reactions, metabolism, fermentation

RUTLEDGE, Douglas B.Sc., Dip.Ed, These 3ᵉᵐᵉ cycle
Institut National Agronomique
Professor
16, Rue Claude Bernard
F - 75005 Paris
Tel: +33 1 4408 1648
Fax: +33 1 4408 1653
E-Mail: rutledge@inapg.inra.fr

Private address
51, Rue des Vinaigriers
F - 75010 Paris
Tel: +33 1 42053178

Fields of interest
Chemometrics, spectroscopy, analytical chemistry, quality, drying, quality assurance, HACCP, thermal processing, nuclear magnetic resonance

TALOU, Thierry Dr.
École Nationale Supérieure de Chimie
Institut National Polytechnique de Toulouse
Project Manager
ENSCT 118 Route de Narbonne
F - 31077 Toulouse Cédex
Tel: +33 5 62885729
Fax: +33 5 62885730

Private address
15, domaine Beauregard
F - 31450 Donneville
Tel: +33 608 802186

Fields of interest
Fruits and vegetables, aroma active compounds, GC, MS, sensory analysis, sensor technology, analytical chemistry, quality, packaging, MAP, storage

Germany

ALDER, Lutz Dr.rer.nat.
Federal Institute for Health Protection of Consumers and Veterinary Medicine
Senior Chemist
Thielallee 88-92
D - 14195 Berlin
Tel: +49 30 8412 3377
Fax: +49 30 8412 3685
E-Mail: l.alder@bgvv.de

Private address
An der Korsopromenade 6
D - 15738 Zeuthen
Tel: +49 33762 70100

Fields of interest
Bread and cereals, dairy, eggs, fish and fish products, fruits and vegetables, legumes, pesticides, GC, HPLC, MS

AUST, Olivier
Heinrich Heine University Düsseldorf
Institut for Physiological Chemistry
PhD-Student
Universitätsstr. 1
D - 40225 Düsseldorf
Tel: +49 211 8112713
Fax: +49 211 8113029
E-Mail: austo@uni-duesseldorf.de

Fields of interest
Antioxidants, carotenoids, proteins, vitamins, HPLC, analytical chemistry, bioavailability, metabolism, nutrition

BALTES, Werner Univ. Prof., Dr.
Techn. University Berlin
Institute for Food Chemistry
Univ. Prof. emeritus
Gustav-Meyer-Allee 25
D - 13355 Berlin
Tel: +49 30 31472819
Fax: +49 30 31472823
E-Mail: werner.baltes@t-online.de

Private address
Schützenallee 8A
D - 14169 Berlin
Tel: +49 30 8141058
Fax: +49 30 8141058

Fields of interest
Coffee, tea, cocoa, aroma active compounds, carbohydrates, Maillard reaction products, heterocyclic aromatic amines, GC, MS, thermal anlaysis, TLC, food composition

BAUER, Ulrich Professor, Dr. rer. nat.
Amt für Umweltschutz und Lebensmitteluntersuchung der Stadt Bonn
Bundesstadt Bonn
Leitender Städt. Chemiedirektor, Amtsleiter
Engeltalstr. 4
D - 53103 Bonn
Tel: +49 228 772351
Fax: +49 228 773956
E-Mail: ulrich.bauer@bonn.de

Private address
Herbstbenden 2
D - 53347 Alfter
Tel: +49 228 649565

Fields of interest
Water, antimicrobials, PCBs, pesticides, toxic trace elements, GC, toxicology

BERGER, Ralf Günter Prof., Dr.
Universität Hannover
Department of Food Chemistry
Institutsleiter
Wunstorfer Str. 14
D - 30453 Hannover
Tel: +49 511 7624581
Fax: +49 511 7624547
E-Mail: lmchemie@mbox.lci.uni-hannover.de

Private address
Karl-Jakob-Hirsch-Weg 33
D - 30455 Hannover
Tel: +49 511 471112

Fields of interest
Beverages, bread and cereals, fruits and vegetables, spices, additives, antioxidants, aroma active compounds, colours, GC, HPLC, MS, PCR, spectroscopy, TLC, analytical chemistry, chemical reactions, enzymology, metabolism, microbiology, quality, biotechnology, fermentation

BETSCHE, Thomas Ph.D.
Bundesanstalt für Getreide-, Kartoffel- und Fettforschung
Head of Institute
Schützenberg 12
D - 32756 Detmold
Tel: +49 5231 741120
Fax: +49 5231 741130
E-Mail: betsche.bagkf@t-online.de

Fields of interest
Bread and cereals, legumes, potatoes, starch, transgenic foods, biopolymers, carbohydrates, carotenoids, dietary fibres, minerals, proteins, trace elements, vitamins, alkaloids, mycotoxins, PCBs, pesticides, plant toxins, toxic trace elements, AAS, analytical chemistry, bioavailability, enzymology, food composition, metabolism, nutrition, quality, regulative issues, fermentation, rheology

BÖHM, Volker Priv.Doz., Dr.rer.nat.habil
University Jena
Institute of Nutrition
Assistant
Dornburger Straße 25 - 29
D - 07743 Jena
Tel: +49 3641 949633
Fax: +49 3641 949632
E-Mail: volker.boehm@rz.uni-jena.de

Private address
Eichenstraße 13
D - 07778 Dorndorf-Steudnitz
Tel: +49 36427 70791

Fields of interest
Coffee, tea, cocoa, fruits and vegetables, antioxidants, carotenoids, polyphenols, vitamins, bioavailability

BROCKMANN, Rainer Dr.
Chemisches Untersuchungsamt Bielefeld
Oststr. 55
D - 33604 Bielefeld
Tel: +49 521 513002
Fax: +49 521 513386

Fields of interest
Eggs, fish and fish products, meat and meat products, poultry, spices, carotenoids, colours, mycotoxins, PCBs, GC, HPLC

BROCKMANN, Anneliese Food Chemist
Chemisches Untersuchungsamt der Stadt Hamm
Managing Director
Sachsenweg 6
D - 59073 Hamm
Tel: +49 2381 178501, 178500
Fax: +49 2381 172253

Fields of interest
Management

BRUNN, Hubertus Dipl.-Chem., Prof., Dr.rer.nat.
Staatliches Medizinal-, Lebensmittel- und Veterinäruntersuchungsamt Mittelhessen
Director of the Institute
Marburgerstr. 54
D - 35396 Giessen
Tel: +49 641 300625
Fax: +49 641 300618
E-Mail: hubertus.e.brunn@ernaehrung.uni-giessen.de

Private address
Hintergasse 8
D - 35325 Muecke
Tel: +49 6400 8639
Fax: +49 6400 6598

Fields of interest
Dairy, eggs, fish and fish products, fruits and vegetables, meat and meat products, poultry, drugs, PCBs, pesticides, analytical chemistry, toxicology

BUDDE, Jürgen Dr. Diplom-Chemiker, Lebensmittelchemiker, Apotheker
Chemisches Untersuchungslabor Dr. Budde
Inhaber
Neckarstraße 14
D - 64283 Darmstadt
Tel: +49 6151 24090

Fax: +49 6151 20475
E-Mail: dr.budde@t.online.de

Private address
Novalisstraße 6
D - 64285 Darmstadt
Tel: +49 6151 494380

Fields of interest
Cosmetics, fruits and vegetables, potatoes, sweeteners, additives, HPLC, spectroscopy, analytical chemistry, microbiology, hygiene

BÜNING-PFAUE, Hans Prof., Dr.
Universität Bonn
Department of Food Science and Food Chemistry
Endenicher Allee 11-13
D - 53115 Bonn
Tel: +49 228 732361
Fax: +49 228 733499
E-Mail: buening@uni-bonn.de

Private address
Auf dem Uhlberg 15
D - 53127 Bonn
Tel: +49 228 256578

Fields of interest
Eggs, fruits and vegetables, potatoes, drugs, chemometrics, GC, HPLC, NIR

CHRISTOPH, Norbert Dr.rer.nat.
Landesuntersuchungsamt für das Gesundheitswesen Nordbayern
Food Chemist
Luitpoldstraße 1
D - 97082 Würzburg
Tel: +49 931 41993167
Fax: +49 931 41993210

Private address
Pfarrer-Fröhlich-Str. 11
D - 97295 Waldbrunn

Fields of interest
Beverages, sugar, honey, sugar alcohols, water, wine, aroma active compounds, PAHs, GC, MS, IRMS, NMR, adulteration, analytical chemistry, fermentation

DAPHI-WEBER, Juliane
Chemisches Untersuchungsamt Bielefeld
Lebensmittelchemikerin
Oststr. 55
D - 33604 Bielefeld
Tel: +49 521 516705
Fax: +49 521 513386

Private address
Siekbreede 62
D - 33649 Bielefeld

Fields of interest
PCBs, allergology, toxicology, indoors air pollution, hazardous products

DAWIHL, Gerd Dr.phil.nat.
Zentrales Institut des Sanitätsdienstes der BW Koblenz
Department of Pharmacy and Food Chemistry
Head of Laboratory of Food Chemistry
Jakob-Kaiser-Str. 6
D - 56076 Koblenz
Tel: +49 261 7802240
Fax: +49 261 7802419

Private address
Horchheimer Höhe 1
D - 56076 Koblenz
Tel: +49 261 71477

Fields of interest
Dairy, meat and meat products, proteins, sensory analysis, analytical chemistry, quality

DECHOW, Arndt
Chemisches Untersuchungsamt Bielefeld
Food Chemist
Oststr. 55
D - 33604 Bielefeld
Tel: +49 521 512661
Fax: +49 521 513386

Private address
Greifswalder Straße 54
D - 33605 Bielefeld
Tel: +49 177 4828276

Fields of interest
Water, HPLC, analytical chemistry

DETTWEILER, Gerd Dr.
Hygiene Institute Hamburg
Head of Laboratory
Marckmannstraße 129A
D - 20539 Hamburg
Tel: +49 40 42837363
Fax: +49 40 42837363
E-Mail: baghi413@mailhub.fhhnet.dbp.de

Private address
Marckmannstraße 162
D - 20539 Hamburg

Fields of interest
Fruits and vegetables, aroma active compounds, pesticides, toxic trace elements, AAS, GC, HPLC, MS, analytical chemistry, toxicology

DIETRICH, Helmut Professor, Dr.
Research Institute Geisenheim

Department of Wine Analysis and Beverage Research
Head of Department
Rüdesheimer Straße 28
D - 65366 Geisenheim
Tel: +49 6722 502311
Fax: +49 6722 502310
E-Mail: h.dietrich@geisenheim.mnd.fh-wiesbaden.de

Private address
Am Tal 5
D - 65385 Rüdesheim
Tel: +49 6726 2627

Fields of interest
Beer, beverages, fruits and vegetables, spirits, sweeteners, wine, antioxidants, biopolymers, carbohydrates, carotenoids, ethanol, polyphenols, AAS, GC, HPLC, MS, sensory analysis, spectroscopy, analytical chemistry, food composition, biotechnology, filtration, minimal processing

---

EBERLE, Mike Dr.
Union Deutsche Lebensmittelwerke GmbH
Head Development Department
Dammtorwall 15
D - 20355 Hamburg
Tel: +49 40 34931973
Fax: +49 40 34931923
E-Mail: mike.eberle@unilever.com

Private address
Wismarring 12
D - 25436 Tomesch

Fields of interest
Beverages, fruits and vegetables, potatoes, starch, food composition, management, high pressure technology, packaging, MAP, preservation, quality assurance, HACCP

EHLERS, Dorothea Prof.Dr.
University of Applied Sciences Berlin
FB V
Professor, Teacher
Kurfürstenstraße 141
D - 10785 Berlin
Tel: +49 30 45044136
Fax: +49 30 45044142
E-Mail: ehlers@tfh-berlin.de

Private address
Schönburgstraße 2
D - 12103 Berlin
Tel: +49 30 7513601

Fields of interest
Oils and fats, spices, aroma active compounds, lipids and fatty acids, mycotoxins, GC, HPLC, spectroscopy

---

EISENBRAND, Gerhard Prof.
Deparmtment of Chemistry
Head of the Division of Food Chemistry and Environmental Toxicology
Erwin-Schrödinger-Str. 52
D - 67663 Kaiserslautern
Tel: +49 631 2052973
Fax: +49 631 2053085
E-Mail: eosenbran@rhok.uni-kl.de

Private address
Gustav-Kirchhoff-Str. 33
D - 69120 Heidelberg
Tel: +49 6221 480089
Fax: +49 6221 475984

Fields of interest
Cosmetics, additives, aroma active compounds, proteins, steroids, alkaloids, hormones, PAHs, Nitrosamines, AAS, GC, HPLC, MS, PCR, TLC, chemical reactions, enzymology, metabolism, toxicology

---

ENDERS, Peter W. Dipl.-Chem.
Springer-Verlag GmbH & Co. KG
Senior Editor
Tiergartenstraße 17
D - 69121 Heidelberg
Tel: +49 6221 487541
Fax: +49 6221 487330
E-Mail: enders@springer.de

Private address
Tel: +49 6203 692093

Fields of interest
Fish and fish products, meat and meat products, poultry, wine, aroma active compounds, carbohydrates, ethanol, pesticides, plant toxins, toxic trace elements, chemometrics, electrochemistry, electrophoresis, GC, HPLC, MS, analytical chemistry, quality, biotechnology, quality assurance, HACCP

ERNING, Dieter Food Chemist
Institute of Chemical Investigation Hamm
Laboratory Manager
Sachsenweg 6
D - 59073 Hamm
Tel: +49 2381 178570
Fax: +49 2381 172253
E-Mail: erning@stadt.hamm.de

Private address
Maximilianstr. 26
D - 59071 Hamm
Tel: +49 2381 889484

Fields of interest
Water, AAS, HPLC, analytical chemistry, microbiology, regulative issues, hygiene, irradiation, quality assurance, HACCP

FRANZKE, Claus em.Prof., Dr.Ing.habil., Dr.sc.nat.
Technische Universität
Inst. f. Lebensmittelchemie
Vorsitzender des Regionalverbandes Nordost der
Lebensmittelchemischen Gesellschaft
Gustav-Meyer-Allee 25
D - 12526 Berlin
Tel: +49 30 31472805

Private address
Jahnstr. 8
D - 12526 Berlin
Tel: +49 30 6766617

Fields of interest
Oils and fats, lipids and fatty acids, biotechnology

GERTZ, Christian Dr.
Institute of Chemical Investigation
Chemisches Untersuchungsamt Hagen
Director
Pappelstr. 1
D - 58099 Hagen
Tel: +49 2331 2074715, 2074726
Fax: +49 2331 2072454

Private address
Schmalenbeckstr. 1C
D - 58093 Hagen
Tel: +49 2334 924277
Fax: +49 2334 924278

Fields of interest
Oils and fats, antioxidants, fat replacers, lipids and fatty acids, oxidation reactions, PAHs, chemometrics, GC, HPLC, sensory analysis, adulteration, analytical chemistry, food composition, quality, quality assurance HACCP, thermal processing

---

GLÜCK, Bernfried Staatl. geprüfter Lebensmittelchemiker
Chemisches und Veterinäruntersuchungsamt, Sigmaringen
Head of Department of Meat and Fish
Fidelis-Graf-Str. 1
D - 72488 Sigmaringen
Tel: +49 7571 732711
Fax: +49 7571 732614

Private address
Weingartenstr. 22
D - 72517 Sigmaringendorf
Tel: +49 7571 5626

Fields of interest
Fish and fish products, meat and meat products, poultry, additives, antioxidants, colours, proteins, biogenic amines, PAHs, AAS, electrophoresis, GC, HPLC, PCR, analytical chemistry, food composition, quality, quality assurance, HACCP

HAFFKE, Helma
Chemisches Untersuchungsamt Bielefeld
Lebensmittelchemikerin, Sachgebietsleitung
Oststr. 55
D - 33604 Bielefeld
Tel: +49 521 512659
Fax: +49 521 513386
E-Mail: gaida.haffke@t-online.de

Private address
Westphalenweg 47a
D - 33104 Paderborn

Fields of interest
Bread and cereals, colours, trace elements, toxic trace elements, AAS, TLC, analytical chemistry, textiles

---

HAHN, Harald Dr.rer.nat
Chemisches und Veterinäruntersuchungsamt
Laborleiter
Hedinger Straße 2A
D - 72448 Sigmaringen
Tel: +49 7571 732623
Fax: +49 7571 732602
E-Mail: hahnh@clua.cvuatu.bwl.de

Private address
Uhlandweg 8
D - 72571 Sigmaringen
Tel: +49 7571 50995

Fields of interest
Additives, aroma active compounds, PAHs, PCBs, GC, MS, analytical chemistry, toxicology

---

HANEWINKEL-MESHKINI, Susanne Dr.rer.nat.
Chemisches Untersuchungsamt Bielefeld
Leader of Laboratory
Oststr. 55
D - 33604 Bielefeld
Tel: +49 521 512660
Fax: +49 521 3386

Fields of interest
Coffee, tea, cocoa, fruits and vegetables, legumes, sugar, honey, sugar alcohols, sweeteners, additives, HPLC

HEINZLER, Matthias Dr.
Untersuchungsamt Nordhessen
Analyst
Druseltalstraße 67
D - 34131 Kassel
Tel: +49 561 3101179
Fax: +49 561 3101242

Private address
Henschelweg 7
D - Ahnatal
Tel: +49 5609 9905

Fields of interest
Fruits and vegetables, potatoes, pesticides, GC, HPLC, MS, analytical chemistry, quality

---

HENLE, Thomas Prof., Dr.
Technical University of Dresden
Institute of Food Chemistry
Head/Director
Mommsenstr. 13
D - 01062 Dresden
Tel: +49 351 4634647
Fax: +49 351 4634138

Fields of interest
Beer, bread and cereals, coffee, tea, cocoa, cosmetics, dairy, sugar, honey, sugar alcohols, transgenic foods, amino acids, antioxidants, biopolymers, Maillard reaction products, oxidation reactions, polyphenols, proteins, biogenic amines, AAS, CE, electrophoresis, GC, HPLC, MS, PCR, spectroscopy, TLC, adulteration, allergology, analytical chemistry, bioavailability, chemical reactions, food composition, metabolism, nutrition, quality, biotechnology, drying, fermentation, high pressure technology, rheology, storage, thermal processing

---

HEY, Hanke Dr.
Official Food Chemistry and Veterinary Test Institute
Head of Institute
Max-Eyth-Straße 5
D - 24537 Neumünster
Tel: +49 4321 904610
Fax: +49 4321 904619
E-Mail: hanke.hey@lvua-sh.de

Private address
Frohauweg 48
D - 24111 Kiel
Tel: +49 431 697047

Fields of interest
Additives, accreditation, validation, quality assurance, HACCP, food law, food law enforcement, food hazards and risk assessment

---

HILS, Arno K. A. Staatl. gepr. Lebensmittelchemiker, European Chemist
Rich. Hengstenberg GmbH u. Co.
Leiter des Zentrallabors
Mettingerstraße 109
D - 73728 Esslingen
Tel: +49 711 3929207
Fax: +49 711 3929448

Fields of interest
Fruits and vegetables, wine, electrochemistry, HPLC, sensor analysis, spectroscopy, analytical chemistry, management, quality, LIMS

HONIKEL, Karl Otto Dr. rer. nat.
Bundesanstalt für Fleischforschung
Institut für Chemie und Physik
Dir. u. Prof. (Institutsleiter)
E.-C.-Baumann.-Str. 20
D - 95326 Kulmbach
Tel: +49 9221 8031
Fax: +49 9221 803303, 803200
E-Mail: kohonikel.baff@t-online.de

Private address
Amselweg 13
D - 95326 Kulmbach

Fields of interest
Meat and meat products, poultry, additives, antioxidants, PAHs, PCBs, toxic trace elements, MS, PCR, sensor technology, food composition

---

JANSON-MUNDEL, Ortrun Dr.rer.nat., EurChem
RWTÜV Anlagentechnik GmbH
TÜV CERT Lead Auditor (ISO 9000, ISO 14001), Umweltgutachterin (EMAS)
Langemarckstraße 20
D - 45141 Essen
Tel: +49 201 8253404
Fax: +49 201 8253248
E-Mail: o.janson-mundel@rwtuev-at.de

Private address
Lothringenstraße 18A
D - 45259 Essen
Tel: +49 201 464093

Fields of interest
Hygiene, quality assurance, HACCP, EMAS

JÖRISSEN, Urban Dr.rer.nat
Handels- und Umweltschutzlaboratorium Dr. Wiertz - Dipl.Chem Eggert - Dr. Jorissen GmbH
Dr. Wiertz-Eggert-Dr. Jörissen GmbH
Managing Director
Stenzelring 14B
D - 21107 Hamburg
Tel: +49 40 75270928
Fax: +49 40 75270935
E-Mail: urban.joerissen@wej.de

Private address
Hochkamp 39
D - 21244 Buchholz
Tel: +49 4181 7748
Fax: +49 4181 281996

Fields of interest
Coffee, tea, cocoa, oils and fats, sugar, honey, sugar alcohols, transgenic foods, mycotoxins, GC, HPLC, adulteration, analytical chemistry, food composition

KLEIN, Erich
Chemisches- und Veterinäruntersuchungsamt
Director
Hedinger Straße 2A
D - 72448 Sigmaringen
Tel: +49 7571 732629, 732605
Fax: +49 7571 732602
E-Mail: klein@clua.cvuatu.bwl.de

Private address
Am Riedbaum 15
D - 72488 Sigmaringen
Tel: +49 7571 14394
Fax: +49 7571 683942

Fields of interest
Meat and meat products, poultry, sugar, honey, sugar alcohols, PAHs, GC, HPLC, analytical chemistry, consumer research, management, quality, quality assurance, HACCP

KLOSTERMEYER, Henning o.Prof., Dr.
Technische Universität München
Chair of Chemistry of Biopolymers
Full Professor
Weihenstephaner Berg 3
D - 85350 Freising
Tel: +49 8161 713500
Fax: +49 8161 714404
E-Mail: biopolymere@lrz.tum.de

Private address
Giggenhauserstr. 16; D - 85354 Freising
Tel: +49 8161 94454

Fields of interest
Dairy, biopolymers, proteins, biogenic amines, allergology, analytical chemistry, chemical reactions, enzymology, immunology, biotechnology, genetic engineering

KNIELING, Ralph G.
Springer-Verlag GmbH & Co. KG
c/o Enders
Hirschberger Str. 6
D - 69198 Schriesheim
Tel: +49 6203 692093

Fields of interest
Meat and meat products, poultry, additives, aroma active compounds, ethanol, lipids and fatty acids, trace elements, vitamins, biogenic amines, pesticides, plant toxins, chemometrics, electrophoresis, GC, HPLC, MS, polarography, spectroscopy, TLC,

analytical chemistry, quality, regulative issues, toxicology, biotechnology, quality assurance, HACCP

KOMBAL, Ralph Dr.rer.nat.
Staatliches Lebensmitteluntersuchungsamt Oldenburg
Arbeitsgruppen- und Fachbereichsleiter
Postfach 2462
D - 26014 Oldenburg
Tel: +49 441 9803163
Fax: +49 441 9803121
E-Mail: 100432.3100@compuserve.com

Private address
Am Gutleuthaus 17
D - 76829 Landau
Tel: +49 6341 60567
Fax: +49 441 384606

Fields of interest
Mycotoxins, PAHs, pesticides, HPLC, MS, TLC, regulative issues, veterinary chemotherapeutics, ELISA

KRAMER, Jörg Dr. rer. nat.
Langnese-Iglo GmbH
Quality Assurance Manager
Aeckern 1
D - 48734 Reken
Tel: +49 2864 82245
Fax: +49 2864 82425

Fields of interest
Bread and cereals, fruits and vegetables, meat and meat products, poultry, analytical chemistry, management, microbiology, quality, freezing, quality assurance, HACCP

KRAUSE, Wolfgang Prof., Dr. rer. nat. habil.
Dresden University of Technology
Department of Food Chemistry
Teacher and scientist
Mommsenstr. 13
D - 01062 Dresden
Tel: +49 351 4635122
Fax: +49 351 4634138

Private address
Industriestr. 46; D - 01129 Dresden

Fields of interest
Dairy, proteins, Enzymes, CE, electrophoresis, HPLC, PCR, analytical chemistry, enzymology, fermentation

KROH, Lothar W. Univ. Prof., Dr.rer.nat.
Technische Universität Berlin
Institute of Food Chemistry
Head of Institute
Gustav-Meyer-Allee 25

D - 13355 Berlin
Tel: +49 30 31472 584
Fax: +49 30 31472 585
E-Mail: lothar.kroh@tu-berlin.de

Private address
Wodanstr. 72
D - 13156 Berlin
Tel: +49 30 9171443

Fields of interest
Starch, carbohydrates, Maillard reaction products, GC, HPLC, MS, TLC, analytical chemistry, chemical reactions, thermal processing

---

KROLL, Jürgen Prof., Dr.rer.nat
Universität Potsdam
Institut für Ernährungswissenschaft
Professur für Lebensmittelchemie
Arthur-Scheunert-Allee 114-116
D - 14558 Berholz-Rehbrücke
Tel: +49 33200 88528
Fax: +49 33200 88573
E-Mail: jkroll@rz.uni-potsdam.de

Private address
Dieselstr. 13
D - 14482 Potsdam
Tel: +49 331 712812

Fields of interest
Biopolymers, polyphenols, proteins, electrophoresis, HPLC, chemical reactions, food composition, nutrition

---

LACH, Günter Dr.rer.nat
Dr. Wiertz - Dipl.Chem. Eggert - Dr. Jörissen GmbH
Head of Department Residues and Contaminants
Stenzelring 14B
D - 21107 Hamburg
Tel: +49 40 7527090, 75270943
Fax: +49 40 75 27 09 35
E-Mail: guenter.lach@wej.de

Private address
Metzendorfer Weg 74
D - 21077 Hamburg
Tel: +49 40 7600005
Fax: +49 40 7600005

Fields of interest
Fruits and vegetables, oils and fats, mycotoxins, PAHs, PCBs, pesticides, toxic trace elements, GC, MS, analytical chemistry

---

LAMBERT, Michael Food Chemisty
Quality Services International GmbH
Managing Director
Flughafendamm 9a

D - 28199 Bremen
Tel: +49 421 594770
Fax: +49 421 594771
E-Mail: info@qsi-q3.de, info@quain.de

Fields of interest
Spices, sugar, honey, sugar alcohols, GC, HPLC, sensory analysis, adulteration, management, quality, regulative issues, quality assurance, HACCP

LANDSIEDEL, Robert Dr.rer.nat
BASF AG
DUP-Z470
D - 67056 Ludwigshafen
Tel: +49 621 6058144
Fax: +49 621 6051734
E-Mail: robert.Landsiedel@basf-ag.de

Private address
Martinsgasse 5; D - 67547 Worms
Tel: +49 6241 412615

Fields of interest
Regulative issues, toxicology, food contact material

LAUTENBACHER, Lutz-Michael Dr.
Sachverständigenbüro Dr. Lautenbacher
Possessor
Landsberger Str. 495
D - 82141 München
Tel: +49 89 82020020, 82020021
Fax: +49 89 82020022

Private address
Landsberger Str. 495
D - 82141 München
Tel: +49 89 82020020
Fax: +49 89 82020022

Fields of interest
Lipids and fatty acids, minerals, trace elements, vitamins, analytical chemistry, consumer research, management, microbiology, regulative issues, toxicology

---

LINDHAUER, Meinolf G. Dr.rer.nat
Institute of Cereal, Potato and Starch Technology
Head of Department, Director, Professor
Schützenberg 12
D - 32756 Detmold
Tel: +49 5231 741420
Fax: +49 5231 741100
E-Mail: lindhauer.bagk@t-online.de

Private address
Billerbecker-Str. 1
D - 32805 Horn - Bad Meinberg
Tel: +49 5233 4885

Fields of interest
Bread and cereals, legumes, potatoes, starch, carbohydrates, carotenoids, dietary fibres, proteins, mycotoxins, quality

---

LITTMANN-NIENSTEDT, Sigrid Food Chemist
Institute of Chemical Investigation
Laboratory Manager
Sachsenweg 6
D - 59073 Hamm
Tel: +49 2381 178510, 178501
Fax: +49 2381 172253

Fields of interest
Eggs, meat and meat products, poultry

LORENZEN, Kay
Pharma Hameln GmbH
Quality assurance/validation
Langes Feld 13
D - 31789 Hameln
Tel: +49 5151 581284
Fax: +49 5151 581258
E-Mail: k.lorenzen@pharma-hameln.de, lorenzen@debitel.net

Private address
In der Korn 20A
D - 31789 Hameln
Tel: +49 5151 924497
Fax: +49 5151 924497

Fields of interest
Quality, quality assurance, HACCP, validation of processes

LUCKAS, Bernd Prof. Dr.habil.
Friedrich-Schiller-Universität
Institut für Ernährungswissenschaften
Ordinarius for Food Science
Dornburger Str. 25
D - 07743 Jena
Tel: +49 3641 949650, 949651
Fax: +49 3641 949352

Private address
Zenkerweg 5
D - 07743 Jena
Tel: +49 3641 448464

Fields of interest
Fish and fish products, meat and meat products, poultry, oils and fats, amino acids, lipids and fatty acids, biogenic amines, marine toxins, mycotoxins, PCBs, pesticides, CE, GC, HPLC, MS, analytical chemistry

---

LUDWIG, Eberhard Prof., Dr. rer. nat. habil., Dipl.Chem., Dipl.-Lebensmittelchemiker
Dresden University of Technology
Institut of Food Chemistry
Professor emeritus
Mommsenstr. 13
D - 01062 Dresden
Tel: +49 351 4634647
Fax: +49 351 4634138

Private address
Regensburger Str. 23
D - 01187 Dresden
Tel: +49 351 4712110

Fields of interest
Coffee, amino acids, proteins, peptides

MAIER, Hans Gerhard Dr.
Institut für Lebensmittelchemie der TU Braunschweig
Professor, retired
Schleinitzstr. 20
D - 38106 Braunschweig
Tel: +49 531 3917202
Fax: +49 531 3917230

Private address
Hermann-Rautmann-Str. 7
D - 38116 Braunschweig
Tel: +49 531 512726

Fields of interest
Coffee, tea, cocoa

MALISCH, Rainer Dr.
Chemisches und Veterinäruntersuchungsamt Freiburg
Head of Dioxin Laboratory, Head of Residue Analysis Department
Bissierstraße 5
D - 79114 Freiburg
Tel: +49 761 88550
Fax: +49 761 8855100
E-Mail: malisch@clva.cvvafr.bwl.de

Private address
Im Laimacker 9
D - 79249 Merzhausen
Tel: +49 761 406641

Fields of interest
Drugs, hormones, PCBs, pesticides, MS, dioxin

MALWITZ, Dietmar Dr.rer.nat., Dipl.-Chem.
Union Deutsche Lebensmittelwerke GmbH
Compyy QA Manager
Dammtorwall 15
D - 20355 Hamburg
Tel: +49 40 34931901
Fax: +49 40 359211901

Germany

E-Mail: dietmar.malwitz@unilever.com

Private address
Basaltweg 16
D - 22395 Hamburg

Fields of interest
Dairy, fish and fish products, meat and meat products, poultry, oils and fats, food composition, management, quality, regulative issues, quality assurance, HACCP

MARX, Friedhelm Dr.rer.nat.
Rheinische Friedrich-Wilhelms-Universität Bonn
Institut für Lebensmittelwissenschaft und Lebensmittelchemie
Akademischer Oberrat, Privatdozent
Endenicher Allee 11-13
D - 53115 Bonn
Tel: +49 228 733713
Fax: +49 228 733757
E-Mail: f.marx@uni-bonn.de

Private address
Niederberger Str. 50
D - 53909 Zülpich
Tel: +49 2252 4662

Fields of interest
Fruits and vegetables, oils and fats, antioxidants, aroma active compounds, lipids and fatty acids, GC, MS, analytical chemistry

---

MATISSEK, Reinhard Prof., Dipl.-Ing., Dr.rer.nat.habil.
Lebensmittelchemisches Institut des BDSI
Director
Adamsstrasse 52-54
D - 51063 Köln
Tel: +49 221 623061
Fax: +49 221 610477
E-Mail: lei-koeln@lci-Koeln.de

Fields of interest
Coffee, tea, cocoa, sugar, honey, sugar alcohols, sweeteners, colours, fat replacers, lipids and fatty acids, alkaloids, GC, HPLC, food composition

---

MISCHNICK, Petra Prof., Dr.
TU Braunschweig
Institut für Lebensmittelchemie
Schleinitzstr. 20
D - 38106 Braunschweig
Tel: +49 531 3917201
Fax: +49 531 3917230
E-Mail: p.mischnick@tu-bs.de

Private address
Gärtnerwinkel 33
D - 38302 Wolfenbüttel

Tel: +49 5331 979101

Fields of interest
Starch, biopolymers, carbohydrates, dietary fibres, CE, GC, MS, analytical chemistry, chemical reactions

---

MÖRSEL, Jörg-Thomas Dr.
Technical University Berlin
Institute for Food Chemistry
Gustav-Meyer-Allee 25
D - 13355 Berlin
Tel: +49 30 31472821
Fax: +49 30 31472823
E-Mail: thomas.moersel@tu-berlin.de

Private address
Altlandsberger Ch. 63a
D - 15370 Fredersdorf
Tel: +49 33439 77243
Fax: +49 33439 77244

Fields of interest
Oils and fats, lipids and fatty acids, vitamins, GC, HPLC, analytical chemistry, chemical reactions, food composition, quality, irradiation

---

MOSANDL, Armin Prof., Dr.
Department of Food Chemistry
Managing Director
Marie-Curie-Str. 9
D - 60439 Frankfurt am Main
Tel: +49 69 798 29203, 29202
Fax: +49 69 798 29207
E-Mail: mosandl@em.uni-frankfurt.de

Private address
Würzburger Str. 13
D - 97337 Dettelbach
Tel: +49 9324 1803
Fax: +49 9324 1803

Fields of interest
Fruits and vegetables, Oils and vegetables, oils and fats, spices, spirits, aroma active compounds, ethanol, lipids and fatty acids, GC, IRMS, adulteration, analytical chemistry, quality assurance, HACCP

NEHRING, Ulrich P. Dr.rer.nat.
Institut Nehring GmbH
Manager
Bismarckstraße 7
D - 38102 Braunschweig
Tel: +49 531 238990
Fax: +49 531 2389977
E-Mail: institut.nehring@t-online.de

Private address
Helmstedter Str. 159

D - 38102 Braunschweig
Tel: +49 531 790707

Fields of interest
Fish and fish products, fruits and vegetables, potatoes, spices, additives, mycotoxins, pesticides, analytical chemistry, regulative issues, preservation

NÖHLE, Ulrich Dr.rer.nat.
Nestlé Deutschland AG
Head of Supply Chain Management
D - 60523 Frankfurt
Tel: +49 69 66713969
Fax: +49 69 66714565
E-Mail: ulrich.noehle@de.nestle.com

Fields of interest
Coffee, tea, cocoa, dairy, oils and fats, sugar, honey, sugar alcohols, additives, management, quality, regulative issues, hygiene, quality assurance, HACCP

OCHS, Stefan Dr.
Seeberger KG
Hans-Lorenser-Str. 36
D - 89079 Ulm
Tel: +49 731 4093180
Fax: +49 71 4093266
E-Mail: s.ochs@seeberger-ulm.de

Private address
Holzbachstr. 39
D - 86152 Augsburg
Tel: +49 821 514474

Fields of interest
Coffee, tea, cocoa, fruits and vegetables, antioxidants, mycotoxins, pesticides, GC, HPLC, quality

OTTENEDER, Herbert Dr.
Institute of Chemical Investigation Trier
Director, Food Chemist
Maximineracht 11A
D - 54295 Trier
Tel: +49 651 1446212, 1446232
Fax: +49 651 21028
E-Mail: cuatrier.otteneder@t-online.de

Fields of interest
Beverages, transgenic foods, water, wine, mycotoxins, pesticides, adulteration

PETZ, Michael Dr.rer.nat, Professor
University of Wuppertal
Department of Food Chemistry
Head of Department
Gauss-Straße 20
D - 42119 Wuppertal
Tel: +49 202 4392783

Fax: +49 202 4392784
E-Mail: petz@uni-wuppertal.de

Fields of interest
Dairy, eggs, meat and meat products, poultry, spices, drugs, chemometrics, GC, HPLC, MS, analytical chemistry

PFALZGRAF, Andreas Dr.
Land Sachsen Anhalt, Landesveterinär- und Lebensmitteluntersuchungsamt
Dezernatsleiter
Freiimfelderstraße 66/68
D - 06112 Halle
Tel: +49 345 5643173
Fax: +49 345 5643439
E-Mail: poststelle@lvluahal.ml.lsa--net.de

Private address
Frittz-Hoffmann Straße 57
D - 06116 Halle
Tel: +49 345 2906572

Fields of interest
PAHs, PCBs, pesticides, analytical chemistry, food composition, quality, regulative issues

PISCHETSRIEDER, Monika Prof., Dr.
Universität Erlangen - Nürnberg
Institut für Pharmazie und Lebensmittelchemie
Professor for Food Chemistry
Schuhstr. 19
D - 91052 Erlangen
Tel: +49 9131 8524102
Fax: +49 9131 8522587
E-Mail: pischetsrieder@lmchemie.uni-erlangen.de

Fields of interest
Dairy, Maillard reaction products, GC, HPLC, MS, analytical chemistry, chemical reactions, toxicology, thermal processing

PLAGA-LODDE, Annette
Chemisches Untersuchungsamt Bielefeld
Oststr. 55
D - 33604 Bielefeld
Tel: +49 521 512658
Fax: +49 521 513386

Fields of interest
Meat and meat products, poultry, immunology, microbiology

POLLMER, Udo Lebensmittelchemiker
European Institute of Food and Nutrition Sciences
Head of Scientific Affairs
Amselweg 7
D - 65239 Hochheim

Tel: +49 6145 970201
Fax: +49 6145 970202

Private address
Eppinger Str. 4
D - 75050 Gemmingen
Tel: +49 7267 911180
Fax: +49 7267 911181

Fields of interest
Additives, fat replacers, vitamins, pesticides, plant toxins, toxic trace elements, adulteration, nutrition, quality, toxicology

POPKEN, Anne M. Dr.rer.nat.,Dipl.Food Chemist
ECKES GRANINI GmbH & Co KG
Ludwig Eckes Allee 6
D - 55264 Nieder Olm
Tel: +49 613 6351310
Fax: +49 613 6352310
E-Mail: anne.popken@eckes-ag.de

Fields of interest
Beverages, fruits and vegetables, antioxidants, Aromactive compounds, carotenoids, dietary fibres, polyphenols, vitamins, GC, analytical chemistry, nutrition

PREUSS, Axel Dr.
Chemical and Veterinary State Laboratory
Director
Sperlichstr. 19
D - 48151 Münster
Tel: +49 251 9821215
Fax: +49 251 9821250
E-Mail: preuss@cvua.nrw.de

Fields of interest
Adulteration, analytical chemistry, food composition, management, microbiology, quality, regulative issues, food law

RAGOTZKY, Klaus Dr. rer. Nat
Union Deutsche Lebensmittelwerke GmbH.
Officer Scientific & Regulatory Affairs
Dammtorwall 15
D - 20355 Hamburg
Tel: +49 40 34931184
Fax: +49 40 34931922
E-Mail: klaus.ragotzky@unilever.com

Private address
Hagedornstraße 29
D - 20149 Hamburg
Tel: +49 40 450639288

Fields of interest
Oils and fats, additives, antioxidants, dietary fibres, fat replacers, lipids and fatty acids, polyphenols, vitamins, biotechnology, minimal processing

RIMKUS, Gerhard G. Dr.
Official Food and Veterinary Institute of Schleswig-Holstein
Head of Department
Max-Eyth-Straße 5
D - 24537 Neumünster
Tel: +49 4321 904658
Fax: +49 4321 904619
E-Mail: gerhard.rimkus@lvna-sh.de

Private address
Eichenallee 27
D - 24536 Neumünster
Tel: +49 4321 31179
Fax: +49 4321 38531

Fields of interest
Dairy, eggs, fish and fish products, meat and meat products, poultry, PCBs, pesticides, GC, MS, analytical chemistry, metabolism

RISTOW, Reinhard Dr., Diplomchemiker
Lebensmittelchemiker
Chemisches Untersuchungsamt
Dienststellenleiter
Nikolaus-von-Weis-Str. 1
D - 67346 Speyer
Tel: +49 6232 652110
Fax: +49 6232 652195

Private address
Albert-Schweitzer-Str. 6a
D - 67346 Speyer
Tel: +49 6232 25111

Fields of interest
Meat and meat products, poultry, wine, pesticides, toxic trace elements, AAS, GC, adulteration, analytical chemistry, food composition, regulative issues

ROHN, Sascha Staatl. Geprüfter Lebensmittelchemiker
University of Potsdam
Institute of Nutritional Science
Ph.D. student
Arthur-Scheunert-Allee 114-116
D - 14558 Bergholz-Rehbrücke
Tel: +49 33200 88525
Fax: +49 33200 88573
E-Mail: rohn@rz.uni-potsdam.de

Private address
Tremsdorfer Weg 12
D - 14558 Bergholz-Rehbrücke
Tel: +49 170 5206492

Fields of interest
Amino acids, biopolymers, polyphenols, proteins, electrophoresis, HPLC, MS, regulative issues, biotechnology

---

RÜDT, Ulrich Dr.
Chemisches und Veterinäruntersuchungsamt Stuttgart
Leiter
Schaflandstr. 3/2
D - 70736 Fellbach
Tel: +49 711 975 2260, 1234
Fax: +49 711 588176

Fields of interest
Analytical chemistry

SCHÄFER, Karola Dr.
Deutsches Wollforschungsinstitut an der RWTH Aachen e.V.
Senior scientist, group leader
Veltmanplatz 8
D - 52062 Aachen
Tel: +49 241 4469117
Fax: +49 241 4469100
E-Mail: schaefer@dwi.rwth-aachen.de

Private address
Bergstraße 46
D - 52062 Aachen
Tel: +49 241 25607

Fields of interest
Cosmetics, amino acids, antioxidants, biopolymers, proteins, HPLC, spectroscopy, TLC, analytical chemistry

---

SCHLETT, Claus Dr.
Gelsenwasse AG
Abteilungsleiter
Willy-Brandt-Allee 26
D - 45891 Gelsenkirchen
Tel: +49 209 708379
Fax: +49 209 708666
E-Mail: schlett@gelsenwasser.de

Private address
Jasminweg 9
D - 45770 Marl - Sinsen
Tel: +49 2365 81654

Fields of interest
Water, aroma active compounds, steroids, hormones, marine toxins, PAHs, PCBs, pesticides, toxic trace elements, AAS, GC, HPLC, MS, PCR, quality

---

SCHMIDT, Heinz Univ. Prof., Dr., Dipl. Chemiker,
Veterinarian
Tiergesundheitsdienst Bayern e.V.
Head of Dept. Food Hygiene, Assistant Manager
Senator-Gerauer-Str. 23
D - 85586 Poing
Tel: +49 89 9091241
Fax: +49 89 9091202
E-Mail: lh@tgd.bayern.de

Private address
Schalldorfer Str. 2
D - 83550 Emmering

Fields of interest
Dairy, eggs, fish and fish products, meat and meat products, poultry, drugs, HPLC, MS, toxicology, hygiene

---

SCHMOLCK, Werner Dr.rer.nat
MUVA Kempten
Qualitäts- und Laborzentrum für Milchprodukte
Director
Hirnbeinstraße 10
D - 87435 Kempten
Tel: +49 831 52900
Fax: +49 831 52 90 100
E-Mail: wschmolck@muva.de

Fields of interest
Dairy, mycotoxins, PCR, sensory analysis, analytical chemistry, management, quality, hygiene, packaging, MAP, quality assurance, HACCP

SCHNEIDER, Rüdiger Dr.rer.nat
Chemisches und Vet.untersuchungsamt
Head of Department, Quality Manager
Weißenburger Str. 3
D - 76187 Karlsruhe
Tel: +49 721 9263602
Fax: +49 721 9263549
E-Mail: schneider@clua.cvuaka.bwl.de

Private address
Ste.-Marie-aux-Mines-Str. 4
D - 76646 Bruchsal
Tel: +49 7257 931230
Fax: +49 7257 931230

Fields of interest
Minerals, trace elements, toxic trace elements, AAS, analytical chemistry, quality

---

SCHREIER, Peter Prof., Dr.
Universität Würzburg
Lehrstuhl für Lebensmittelchemie
Chair of Food Chemistry
Am Hubland
D - 97074 Würzburg
Tel: +49 931 888 5481
Fax: +49 931 888 5484
E-Mail: schreier@pzlc.uni-wuerzburg.de

Private address
Hans-Gebhardt-Str. 36
D - 97280 Remlingen
Tel: +49 9369 2405
Fax: +49 9369 8167

Fields of interest
Beverages, fruits and vegetables, aroma active compounds, GC, HPLC, MS, analytical chemistry, chemical reactions, microbiology, biotechnology

SCHRENK, Dieter Dr. med., Dr. rer. nat.
University of Kaiserslautern
Food Chemistry and Environmental Toxicology
Professor
Erwin-Schrödinger-Straße 52
D - 67663 Kaiserslautern
Tel: +49 631 2053043
Fax: +49 631 2054398
E-Mail: schrenk@rhrk.uni-kl.de

Private address
Konrad-Adenauer-Str. 86
D - 67663 Kaiserslautern
Tel: +49 631 3103809

Fields of interest
Drugs, hormones, PAHs, PCBs, pesticides, metabolism, nutrition, regulative issues, toxicology

SCHULTE, Erhard Dr.
University of Münster
Institute of Food Chemistry
Scientific Collaborator
Corrensstr. 45
D - 48149 Münster
Tel: +49 251 833397
Fax: +49 251 833396
E-Mail: eschult@uni-muenster.de

Private address
Soetenkamp 4
D - 48149 Münster
Tel: +49 251 82909

Fields of interest
Fruits and vegetables, oils and fats, lipids and fatty acids, GC, HPLC, food composition

SCHWACK, Wolfgang Dr., Professor
University of Hohenheim
Institute of Food Chemistry
Director
Garbenstr. 28
D - 70599 Stuttgart
Tel: +49 711 4593979
Fax: +49 711 4594096
E-Mail: wschwack@uni-hohenheim.de

Private address
3, Rue du Tilleul
F - 67470 Croettwiller
Tel: +333 88 947989
Fax: +333 88 947989

Fields of interest
Cosmetics, fruits and vegetables, oils and fats, carotenoids, lipids and fatty acids, vitamins, pesticides, analytical chemistry, bioavailability, chemical reactions, food composition

SCHWENKE, Klaus Dieter Prof., Dr.Sc.nat., Dr.rer.nat.
Institute of Applied Protein Chemistry
Leader of Research Group Plant Protein Chemistry
c/o BBA, Stahnsdorfer Damm 81
D - 14532 Kleinmachnow
Tel: +44 33203 305830
Fax: +44 33203 305831
E-Mail: food.proteins@prochem.pm.shuttle.de

Private address
Klaus-Groth-Str. 1
D - 14513 Teltow
Tel: +49 3328 41371

Fields of interest
Legumes, biopolymers, proteins, electrophoresis, HPLC, chemical reactions, food composition, rheology, functional properties

SEIBEL, Wilfried MS in Agriculture, Ph.D. in Agricultural Chemistry
Federal Centre for Cereal, Potato and Lipid Research
Former Head of the Federal Research Centre
Schützenberg 12
D - 32756 Detmold
Tel: +49 5231 741659
Fax: +49 5231 20505

Private address
Alter Postweg 19
D - 32756 Detmold
Tel: +49 5231 38343

Fields of interest
Bread and cereals, potatoes, starch, additives, dietary fibres, mycotoxins, pesticides, nutrition, biotechnology

SIEBERS, Johannes Dr.
Biologische Bundesanstalt für Land- und Forstwirtschaft
Head of Laboratory
Messeweg 11-12
D - 38104 Braunschweig
Tel: +49 531 2993511
Fax: +49 53 2993004
E-Mail: j.siebers@bha.de

Fields of interest
Bread and cereals, fruits and vegetables, legumes, water, pesticides, GC, MS, analytical chemistry, regulative issues

---

SPEER, Karl Prof. Dr.
Dresden University of Technology
Institute of Food Chemistry
Professor
Mommsenstr. 13
D - 01062 Dresden
Tel: +49 351 4633132
Fax: +49 351 4634138
E-Mail: karl.speer@chemie.tu-dresden.de

Fields of interest
Coffee, tea, cocoa, sugar, honey, sugar alcohols, PAHs, pesticides, GC, HPLC, MS

STEINHART, Hans Dr.rer.nat., Dr.agr.habil.
University of Hamburg
Department of Biochemistry and Food Chemistry
Director
Grindelallee 117
D - 20146 Hamburg
Tel: +49 40 428384356
Fax: +49 40 428384342
E-Mail: steinhart@lc.chemie.uni-hamburg.de

Private address
Lerchenweg 15
D - 21244 Buchholz/Nordheide
Tel: +49 4181 32662

Fields of interest
Coffee, tea, cocoa, meat and meat products, poultry, antioxidants, carbohydrates, dietary fibres, lipids and fatty acids, Maillard reaction products, HPLC, MS, spectroscopy

---

STELZ, Alice Dr.oec.troph.
Staatliches Medizinal-, Lebensmittel- und Veterinäruntersuchungsamt Mittelhessen
Leader of laboratory, Vice Leader of Department
Marburgerstrasse 54
D - 35396 Giessen
Tel: +49 641 30060
Fax: +49 641 300618

Private address
Georg-Frank-Strasse 19
D - 35423 Lich
Tel: +49 6404 1026

Fields of interest
Beverages, water, minerals, trace elements, toxic trace elements, AAS, microbiology, nutrition, toxicology, hygiene

---

TAUSCHER, Bernhard Dr.rer.nat.habil., Prof.
Bundesforschungsanstalt für Ernährung
Department of Chemistry and Biology
Director and Professor
Haid- und Neustr. 9
D - 76131 Karlsruhe
Tel: +49 721 6625500
Fax: +49 721 6625503
E-Mail: bernhard.tauscher@bfe.uni-karlsruhe.de

Fields of interest
Fruits and vegetables, legumes, water, additives, amino acids, antioxidants, aroma active compounds, biopolymers, carbohydrates, carotenoids, colours, oxidation reactions, proteins, vitamins, alkaloids, electrophoresis, GC, HPLC, MS, sensor analysis, spectroscopy, analytical chemistry, chemical reactions, quality, high pressure technology, minimal processing, packaging, MAP

TERNES, Waldemar Prof., Dr.
School of Veterinary Medincine Hanover
Chemical Institute
Leader of the working group Chemical Analysis
Bischofsholer Damm 15
D - 30173 Hannover
Tel: +49 511 8567544
Fax: +49 511 8567674
E-Mail: wternes@chemie.tiho-hannover.de

Private address
Alte Dorfstr. 15
D - 29690 Schwarmstedt
Tel: +49 5071 2853
Fax: +49 5071 2853

Fields of interest
Eggs, oils and fats, spices, additives, antioxidants, lipids and fatty acids, polyphenols, drugs, PCBs, AAS, GC, HPLC, MS, Polarography, spectroscopy, TLC, analytical chemistry, food composition, toxicology, high pressure technology, thermal processing

---

THIER, Hans-Peter Prof., Dr.
University of Münster
Institute of Food Chemistry
Head of the Institute
Corrensstr. 45
D - 48149 Münster
Tel: +49 251 8333391
Fax: +49 251 8333396
E-Mail: lc@uni-muenster.de

Fields of interest
Fruits and vegetables, additives, carbohydrates, lipids and fatty acids, PCBs, pesticides, analytical chemistry, chemical reactions, food composition, quality assurance

Germany

VIETHS, Stefan Dr., Priv.-Doz.
Paul Ehrlich - Institut
Head of Section 'Development and Standardization of Allergen Extracts'
Paul-Ehrlich-Str. 51-59
D - 63225 Langen
Tel: +49 6103 772256
Fax: +49 6103 771258
E-Mail: viest@pei.de

Fields of interest
Fruits and vegetables, spices, proteins, electrophoresis, HPLC, PCR, allergology, immunology, biotechnology, genetic engineering

---

VON RYMON LIPINSKI, Gert-Wolfhard Prof., Dr.
Nutrinova Nutrition Specialtics & Food Ingredients GmbH
Director Scientific & Regulatory Affairs
Industriepark Höchst
D - 65926 Frankfurt
Tel: +49 69 3053569
Fax: +49 69 30 516306

Private address
Schlesienstraße 62
D - 65824 Schwalbach
Tel: +49 6196 82497
Fax: +49 6196 82497

Fields of interest
Sweeteners, additives, regulative issues

WEDER, Jürgen, K.P. Apl. Prof., Dr.-Ing., Dipl.-Ing.
Technische Universität München
Department of Food Chemistry
Academic Director
Lichtenbergstr. 4
D - 85748 Garching
Tel: +49 89 2891 3265
Fax: +49 89 2891 4183
E-Mail: lebensmittelchemie@lrz.tum.de

Private address
Birkengrund 2c
D - 85276 Pfaffenhofen

Fields of interest
Legumes, proteins, plant toxins, electrophoresis, PCR, analytical chemistry, enzymology, food composition, nutrition

---

WINKELER, Heinz-Dieter Dr.
Chemisches Untersuchungsamt Bielefeld
Oststr. 55; D - 33604 Bielefeld
Tel: +49 521 513095

Fields of interest
Fruits and vegetables, water, colours, heterocyclic aromatic amines, PAHs, PCBs, pesticides, GC, HPLC, MS, spectroscopy, analytical chemistry

---

WINTERHALTER, Peter Dr.habil.
Universität Braunschweig
Institute of Pharmacy and Food Chemistry
Professor
Schleinitzstr. 20
D - 38106 Braunschweig
Tel: +49 531 3917200
Fax: +49 531 3917230
E-Mail: p.winterhalter@tu-bs.de

Fields of interest
Coffee, tea, cocoa, fruits and vegetables, wine, antioxidants, aroma active compounds, carotenoids, colours

---

WITTENSCHLÄGER, Lutz Dr.
Consusltant for the Food and Packaging Industry
Hardenbergstr. 1
D - 32479 Hille
Tel: +49 5764 512199
Fax: +49 5734 512198
E-Mail: foodinfo@gmx.de

Private address
Hardenbergstr. 1
D - 32479 Hille
Tel: +49 5734 1253

Fields of interest
Management, regulative issues, packaging, MAP, quality assurance, HACCP

WUCHERPFENNIG, Karl Prof., Dr.habil.
Baumann-Gonser-Foundation
Board of Directors
Rüdesheimerstr. 28
D - 65366 Geisenheim
Tel: +49 6722 64937
Fax: +49 6722 502310

Private address
Riederbergstr. 81
D - 65195 Wiesbaden
Tel: +49 611 529215
Fax: +49 611 597595

Fields of interest
Beverages, wine, additives, toxic trace elements, sensory analysis, enzymology, filtration, preservation, thermal processing, fruit juice technology

Greece

ADAMANTIADOU, Sophia Honours Degree on Chemistry
General Chemical State Laboratory - Chemical Service of Kalamata
Chemist
Byronos 50
GR - 24100 Kalamata
Tel: +30 721 80347
Fax: +30 721 80347

Private address
Hpirou
GR - 24100 Kalamata
Tel: +30 721 80112

Fields of interest
Oils and fats, spirits, water, GC, HPLC, spectroscopy, TLC, analytical chemistry, regulative issues

ASSIMAKOPOULOU, Angelique
Chemical State Laboratory
Director General of State Laboratory
A. Tsocha Str. 16
GR - 11521 Athens
Tel: +30 1 6464360
Fax: +30 1 6465123
E-Mail: gxk-personnel@ath.forthnet.gr

Private address
60, Doukissis Plakentias St.
GR - 15127 Melissia Attikis
Tel: +30 8044271

Fields of interest
Transgenic food, adulteration, consumer research, food composition, quality, regulative issues, labelling

BOSKOU, Dimitrius Ph.D., Dr., Prof.
Aristotele University of Thessaloniki
School of Chemistry, Lab. of Food Chemistry
Director of Laboratory
GR - 54006 Thessaloniki
Tel: +30 31 997791
Fax: +30 31 997779
E-Mail: boskou@chem.auth.gr

Private address
Kidonion Str. 14
GR - 54006 Thessaloniki
Tel: +30 31 411478

Fields of interest
Oils and fats, antioxidants, lipids and fatty acids, polyphenols, steroids, adulteration, food composition, regulative issues

CHRONEOS, Ioannis B.Sc., M.Sc., Ph.D.
General Chemical State Laboratory
32, Akti Kondyli + Etolikov
GR - 18510 Pireas
Tel: +30 46 13992
Fax: +30 46 12702

Private address
8, Kraterou Str.
GR - 15771 Athens
Tel: +30 1 7717355

Fields of interest
Dairy, additives, proteins, electrophoresis, GC, HPLC, adulteration

CHRYSAFIDIS, Dimitrios Ph.D.
General Chemical State Laboratory
Chemist
16 Tsoha Street
GR - 11521 Athens
Tel: +30 301 6479000
Fax: +30 301 6465123

Private address
Gerostanti St. 14
GR - 15451 Athens
Tel: +30 1 6439707

Fields of interest
Beverages, coffee, tea, cocoa, sugar, honey, sugar alcohols, sweeteners, additives, carbohydrates, colours, vitamins, HPLC, adulteration, polyphenols

DEMOPOLOUS, Constantinos A. Ph.D.
National & Kapodistrian University of Athen
Professor of Biochemie and Food Chemistry
Panepistimioupolis
GR - 15771 Athens
Tel: +30 1 7274265
Fax: +30 1 7274265
E-Mail: demop@ath.forthnet.gr

Private address
39 Anafis
GR - 11364 Athens
Tel: +30 1 8648963

Fields of interest
Fish and fish products, oils and fats, wine, lipids and fatty acids, HPLC, MS, TLC, allergology, enzymology, nutrition

ECONOMIDES, Anna B.Sc. M.Sc.
TÜV Hellas, QA Department
E.Giavasi 46
GR - 15341 A. Paraskevi
Tel: +30 1 6540195
Fax: +30 1 6528025

E-Mail: fedon@compulink.gr

Private address
Amaryllidos 3, P. Psychiko
GR - 15452 Athens
Tel: +30 1 6723545
Fax: +30 1 8826500

Fields of interest
Bread and cereals, meat and meat products, poultry, hygiene, quality assurance, HACCP, ISO 9000

KOMAITIS, Michael Ph.D., Professor
Agricultural University of Athens
Professor of Food Chemistry
Iera Odos 75
GR - 11855 Athens
Tel: +30 1 5294681
Fax: +30 1 5294681
E-Mail: achem@auadec.aua.gr

Private address
Tsimiski 45
GR - 11473 Athens
Tel: +30 1 3804851

Fields of interest
Oils and fats, antioxidants, aroma active compounds, lipids and fatty acids, GC, HPLC, MS, adulteration, analytical chemistry, food composition

KONTOMINAS, Michael B.S., M.S., Ph.D.
University of Ioannina
Lab. of Food Chemistry and Technology
Deptartment of Chemistry
Professor of Food Chemistry
GR - 45110 Ioannina
Tel: +30 651 98342
Fax: +30 651 44836
E-Mail: mkontomi@cc.uoi.gr

Private address
A' Parodos Panepistimiou
GR - 45110 Ioannina
Tel: +30 651 49027
Fax: +30 651 49027

Fields of interest
Meat and meat products, poultry, water, Lipies and fatty acids, GC, HPLC, MS, sensory analysis, spectroscopy, thermal anlaysis, analytical chemistry, quality, irradiation, packaging, MAP, thermal processing

MARIOLEAS, Panagiotis Degree in Chemistry
General Chemical State Laboratory
Chemist
16 Tsoha Street
GR - 11521 Athens

Tel: +30 301 6479341
Fax: +30 301 6469755
E-Mail: gxk-dxy@ath.forthnet.gr

Private address
33, Patiarchou Ioakim Str.
GR - 10675 Athens
Tel: +30 1 7236207

Fields of interest
Fruits and vegetables, potatoes, spices, sweeteners, additives, GC, HPLC, adulteration, food composition

SAKELLARIOU, Christina Degree in Chemistry
General Chemical State Laboratory
Chemist
16 Tsoha Street
GR - 11521 Athens
Tel: +30 301 6479331
Fax: +30 301 6465123
E-Mail: gxk-dxy@ath.forthnet.gr

Private address
Ipsilantou 34
GR - 15451 Athens
Tel: +30 1 6711509

Fields of interest
Beverages, bread and cereals, coffee Tea Cocoa, fruits and vegetables, sweeteners, additives, HPLC, adulteration, food composition, quality

SCORDAKI, Alexandra Hon.Deg. on Chemistry, Ph.D.
General Chemical State Laboratory - Chemical Service of Kalamata
Chemist
Byronos 50
GR - 24100 Kalamata
Tel: +30 721 80347
Fax: +30 721 80347

Private address
Deak Building Nr.1
GR - 24100 Kalamata
Tel: +30 721 96159
Fax: +30 721 96986

Fields of interest
Oils and fats, sugar, additives, GC, HPLC, spectroscopy, TLC, adulteration, regulative issues

TZIA, Constantina Dr.
National Technical University of Athen
Department of Chemical Engineering, Laboratory of Food Science and Technology
Education, Research
5/2004 Polytechniou St., Polytechnioupoli, Zografou
GR - 15780 Athens
Tel: +30 1 7723165

Fax: +30 1 7723163
E-Mail: tzia@orfeus.chemeng.ntua.gr

Private address
Theatrou 24-28
GR - 18534 Piraeus
Tel: +30 1 4129204
Fax: +30 1 4129204

Hungary

FARKAS, Joszef Prof., Ph.D., D.Sc., Corresp. Memb. Hung. Acad. Sci.
St. Stephens University, Gödöllö
Faculty of Food Science and - Engineering
Professor of Food Science
Villanyi Ut 29-43
H - 1118 Budapest
Tel: +36 1 3726303
Fax: +36 1 3726321
E-Mail: huto@hoya.kee.hu

Private address
Ady Endre Ut 19
H - 1221 Budapest
Tel: +36 1 2260842

Fields of interest
Fruits and vegetables, meat and meat products, poultry, spices, antimicrobials, oxidation reactions, thermal anlaysis, microbiology, high pressure technology, irradiation, preservation

GYÖRI, Zoltán D.Sc.
Debrecen Agricultural University
Department of Food Processing and Quality Testing
Vice-Rector, Head of Department
Böszörményi Str. 138
H - 4032 Debrecen
Tel: +36 52 417752
Fax: +36 52 417572
E-Mail: ggyori@fs2.date.hu

Private address
Tessedik Str. 106
H - 4032 Debrecen
Tel: +36 52 342837
Fax: +36 52 342837

Fields of interest
Bread and cereals, legumes, amino acids, minerals, trace elements, toxic trace elements, AAS, spectroscopy, analytical chemistry, quality assurance

HORVÁTH-MOSONYI, Magda Dr.techn.
Semmelweis University
Faculty of Health Sciences, Department of Dietetics
Associate Professor
Horánszky u 24.
H - 1085 Budapest
Tel: +36 1 1382 660
Fax: +36 1 1382043

Private address
Tusnádi u 45
H - 1225 Budapest
Tel: +36 1 3561680
Fax: +36 1 3561680

Fields of interest
Bread and cereals, fruits and vegetables, dietary fibres, bioavailability, food composition, nutrition

KOVACS, Elisabeth Teresia Ph.D.
JATE University
Szeged College of Food Industry
Associate Professor
Mars ter 7
H - 6724 Szeged
Tel: +36 62 546022
Fax: +36 62 546005
E-Mail: elisabet@szef.u-szeged.hu

Private address
Bihari u. 29A.1.4.
H - 6723 Szeged
Tel: +36 62 492591
Fax: +36 62 492591

Fields of interest
Legumes, additives, electrophoresis, food composition, nutrition, emulsifiers

LASZTITY, Radomir M.Sc., Ph.D., D.Sc.
Technical University of Budapest
Department of Biochemistry and Food Technology
University Professor
Müegyetem rkp.3
H - 1502 Budapest, P.O. Box 91
Tel: +36 1 4631627, +36 1 4631255
Fax: +36 1 4633855
E-Mail: lasztity@ch.bme.hu

Private address
Berend ut. 6
H - 1035 Budapest
Tel: +36 1 885700

Fields of interest
Bread and cereals, legumes, transgenic food, proteins, mycotoxins, food composition, nutrition, quality, minimal processing

LUGASI, Andrea Dr.
National Institute of Food-Hygiene and Nutrition

Hungary

Gyàli ùt 3/a
H - 1097 Budapest
Tel: +36 1 2154130/183
Fax: +36 1 2151545
E-Mail: h8649rod@ella.hu

Private address
Birò U.u.16
H - 1203 Budapest
Tel: +36 1 2841135

Fields of interest
Fruits and vegetables, meat and meat products, poultry, spices, wine, antioxidants, oxidation reactions, polyphenols, HPLC, analytical chemistry, nutrition

MOLNAR, Pal Prof., Dr.
Central Food Research Institute
Scientific Advisor, Director of Food Quality Centre
Herman-Otto-Str. 15
H - 1022 Budapest
Tel: +36 1 1356 5082
Fax: +36 1 2741005, 3558244175
E-Mail: p.molnar@cfri.hu

Private address
Hüvösvölgyi ut 157
H - 1021 Budapest
Tel: +36 1 2005597

Fields of interest
Beverages, fruits and vegetables, sensory analysis, quality, regulative issues, quality assurance, HACCP, storage

MOLNAR-PERL, Ibolya Ph.D., D.Sc.
L. Etovos University
Department of Inorganic and Analytical Chemistry
Full Professor
P.O. Box 32
H - 1518 Budapest 112
Tel: +36 1 2090608
Fax: +36 1 2090602
E-Mail: perlne@para.elte.chem.hu

Private address
Dobsinai street 11
H - 1124 Budapest
Tel: +36 1 3562185

Fields of interest
Fruits and vegetables, sugar, honey, sugar alcohols, amino acids, carbohydrates, GC, HPLC, MS, food composition, carboxylic acids

SALGÒ, Andràs Dr. habil.
Technical University Budapest
Department of Biochemistry and Food Technology
Head of Department
Müegyetem RKP. 3
H - 1111 Budapest
Tel: +36 1 4633854
Fax: +36 1 4633855
E-Mail: salgo@chem.bme.hu

Private address
Görgényi Str. 6/a
H - 1025 Budapest
Tel: +36 1 2751354

Fields of interest
Bread and cereals, biopolymers, proteins, chemometrics, spectroscopy, analytical chemistry, enzymology, food composition, rheology

SASS-KISS, Agnes Ph.D.
Central Food Research Institute
Chemist
Herman-Otto-Str. 15
H - 1022 Budapest
Tel: +36 1 3558838
Fax: +36 1 2741005
E-Mail: a.sass@cfri.hu

Private address
Batthyany Street 1/B
H - 2092 Budakeszi
Tel: +36 23 450039

Fields of interest
Beverages, fruits and vegetables, sweeteners, wine, additives, proteins, biogenic amines, electrophoresis, HPLC, adulteration, analytical chemistry

SIMON-SARKADI, Livia Ph.D.
Technical University of Budapest
Department of Biochemistry and Food Technology
Associate Professor
Müegyetem rkp.3, P.O. Box 91
H - 1502 Budapest
Tel: +36 1 4633862
Fax: +36 1 4633855
E-Mail: sarkadi.bet@chem.bme.hu

Private address
Akacvirag u.10 fsz.2
H - 1173 Budapest
Tel: +36 1 2560868

Fields of interest
Bread and cereals, meat and meat products, poultry, wine, amino acids, proteins, biogenic amines, HPLC, TLC, analytical chemistry, food composition

SZILÁGYI, Szilárd
Debrecen Agricultural University
Department of Food Processing and Quality Testing

Assistant Lecturer
Böszörményi Str. 138
H - 4032 Debrecen
Tel: +36 52 2478888076
Fax: +36 52 417572
E-Mail: szszilgyi@fs2.date.hu

Private address
Vásártér 17
H - 2921 Komárom
Tel: +36 34 346658

Fields of interest
Bread and cereals, fruits and vegetables, legumes, potatoes, amino acids, dietary fibres, proteins, vitamins, mycotoxins, HPLC, analytical chemistry, food composition, quality, toxicology, quality assurance

TÖMÖSKÖZI, Sándor Dr.
Technical University of Budapest
Department of Biochemistry and Food Technology
Ass.Prof.
Müegyetem rkp.3
H - 1111 Budapest
Tel: +36 1 4631419
Fax: +36 1 4633855

Fields of interest
Bread and cereals, dairy, legumes, dietary fibres, proteins, steroids, spectroscopy, analytical chemistry, food composition, nutrition

TÓTH-MARKUS, Marianna Ph.D.
Central Food Research Institute
Chemist
Herman-Otto-Str. 15
H - 1022 Budapest
Tel: +36 1 3558838
Fax: +36 1 2741005
E-Mail: m.toth@cfri.hu

Private address
Ujvidèk tèr 7
H - 1145 Budapest
Tel: +36 1 3832 283

Fields of interest
Beverages, fruits and vegetables, oils and fats, aroma active compounds, GC, MS, adulteration, analytical chemistry, food composition

VARADI, Maria Ph.D.
Central Food Research Institute
Scientific Deputy Director
Herman-Otto-Str. 15
H - 1022 Budapest
Tel: +36 1 3558982

Fax: +36 1 2129853

Private address
Lukacs Gy.St. VII62
H - 1039 Budapest
Tel: +36 1 2432521

Fields of interest
Beer, amino acids, electrochemistry, PCR, sensor technology, spectroscopy, analytical chemistry, food composition, quality, quality assurance, HACCP

VISZKOK, Ferenc
BME Biokemia T&Z
Elemiszertechnikus
H - 1111 Budapest
Tel: +36 1 4631153
E-Mail: viszkok@chem.bme.hu

Fields of interest
Bread and cereals, rheology

Ireland

BYRNE, Briege Eileen B.Sc., Ph.D.
National Food Centre
Research Officer
Dunsiner, Castleknock
Dublin 15
Tel: +353 1 8059500
E-Mail: briegebyrne@ireland.com

Private address
Apartment 23, Whitworth Hall, Morning Star Avenue
Dublin 7

Fields of interest
Beer, meat and meat products, poultry, sweeteners, sensory analysis

HOOD, Ted D.E. B.Sc., Ph.D.
D.E Hood Associates
M.D.
13 Taney Road
Dublin 14
Tel: +343 1 2964194
E-Mail: hoodases@iol.ie

Private address
13 Taney Road
Dublin 14
Tel: +343 1 2964194

Fields of interest
Meat and meat products, poultry, water, food composition, microbiology, quality, biotechnology, freezing, packaging, MAP, storage

MC DONALD, Mark MICI, MRSC, CCHEM, MIFST, MIFSTI, B.Sc., Ph.D.
Department of Agriculture, Food and Rural Development, Central Meat Control Laboratory
Chemist
Abbotstown
Dublin 15
Tel: +353 1 6072606
Fax: +353 1 8214966
E-Mail: cmclda@indigo.ie, markmcdonald@tinet.ie

Private address
60 Castle Riada Drive
Co. Dublin, Castle Road Lulan
Tel: +353 1 6212303

Fields of interest
Meat and meat products, poultry, antimicrobials, steroids, hormones, GC, HPLC, MS, analytical chemistry, quality, regulative issues

ROOS, Yrjö Henrik M.Sc., Ph.D.
University College, Cork
Department of Food Science, Technology and Nutrition
Professor of Food Technology
CORK
Tel: +353 21 902386
Fax: +353 21 270213
E-Mail: yrjo.roos@ucc.ie

Fields of interest
Fruits and vegetables, starch, biopolymers, carbohydrates, thermal anlaysis, chemical reactions, food composition, drying, freezing, rheology

SPILLANE, William J. Ph.D., D.Sc.
National University of Irland
Chemistry Department
Galway
Tel: +353 91 750428
Fax: +353 91 525700
E-Mail: william.spillane@ucg.ie

Fields of interest
Sweeteners

TROY, Declan J. M.Sc., C.Chem.
TEAGASC - The National Food Centre
Food Research
Dublin
Tel: +353 1 8059500
Fax: +353 1 8059550
E-Mail: d.troy@nfc.teagasc.ie

Fields of interest
Meat and meat products, poultry, fat replacers, proteins, electrophoresis, sensory analysis, enzymology, quality, high pressure technology, minimal processing

Italy

ACCORSI, Carla Alberta Professore Associato
University of Ferrara
Department of Chemistry
Researcher
L. Borsari 46
I - 44100 Ferrara
Tel: +39 532 291120
Fax: +39 532 291170
E-Mail: acc@unife.it

Private address
Palestro 23
I - 44100 Ferrara
Tel: +39 532 248623
Fax: +39 532 291170

Fields of interest
Sugar, honey, sugar alcohols, carbohydrates, AAS, HPLC, analytical chemistry, fermentation

ANKLAM, Elke Prof., Dr.
Commission of the European Union, Joint Research Centre Ispra, Institute for Health and Consumer Protection, Food Products and Consumer Goods nit
Unit Head
Via Fermi, TP 260
I - 21020 Ispra
Tel: +39 332 785930
Fax: +39 332 785930

Private address
Via Piave 16
I - 21020 Brebbia
Tel: +39 332 773862

Fields of interest
Beverages, coffee, tea, cocoa, dairy, oils and fats, sugar, honey, sugar alcohols, transgenic foods, additives, antioxidants, lipids and fatty acids, polyphenols, vitamins, alkaloids, biogenic amines, mycotoxins, PCBs, pesticides, plant toxins, CE, GC, HPLC, MS, PCR, spectroscopy, TLC, adulteration, analytical chemistry, management, quality, regulative issues, genetic engineering, packaging, MAP

BERARDO, Nicola Dr.
Istituto Sperimentale per la Cerealicoltura - SOP Bergamo
Researcher
Via Stezzano 24
I - 24126 Bergamo

Tel: +36 035 313132
Fax: +31 035 316054
E-Mail: isc2@spm.it

Fields of interest
Bread and cereals, starch, amino acids, carbohydrates, proteins, mycotoxins, chemometrics, electrophoresis, GC, HPLC, MS, spectroscopy, quality, genetic engineering

BONTEMPELLI, Gino Dr., Prof.
University of Udine
Department of Chemical Sciences and Technology
Full Professor of Analyitcal Chemistry
Via Cotonificio 108
I - 33100 Udine
Tel: +39 432 558842
Fax: +39 432 558803
E-Mail: gino.bontempelli@dstc.uniud.it

Private address
Via De Besi 13
I - 35100 Padova
Tel: +39 49 772484

Fields of interest
Fruits and vegetables, water, wine, antioxidants, oxidation reactions, trace elements, electrochemistry, analytical chemistry

BOTRÈ, Claudio Full Professor
University of Rome "La Sapienza"
Department of Pharmacology and General Physiology
Full Professor of Physical Chemistry
P.le Aldo Moro 5
I - 00185 Roma
Tel: +39 6 49913225
Fax: +39 6 49913888

Private address
Via Emanuele Filiberto 190
I - 00185 Roma
Tel: +39 6 7002560
Fax: +39 6 7002560

Fields of interest
Maillard reaction Products, PAHs, PCBs, toxicology

BOTRÈ, Francesco Dr.
University of Rome "La Sapienza"
Department CGMIA
Assistant Professor
Via del Castro Laurenziano 9
I - 00161 Roma
Tel: +39 06 4976 6217
Fax: +39 06 4452251

Private address
Via Emanuele Filiberto 190

I - 00185 Roma
Tel: +39 06 77205652
Fax: +39 06 23310228

Fields of interest
Fish and fish products, meat and meat products, poultry, wine, polyphenols, steroids, marine toxins, electrochemistry, sensory analysis, toxicology

BURINI, Giovanni Associate Professor
University of Perugia
Department of Food and Nutrition Science
San Costanzo
I - 06126 Perugia
Tel: +39 75 5853905
Fax: +39 75 5853904
E-Mail: gburini@unipg.it

Private address
Marsala 6
I - 06128 Perugia
Tel: +39 75 5054220

Fields of interest
Beverages, dairy, additives, antioxidants, lipids and fatty acids, trace elements, vitamins, toxic trace elements, AAS, chemometrics, GC, HPLC, adulteration, analytical chemistry, nutrition, toxicology

CALZOLARI, Claudio Professore Emerito (University of Trieste), Dr.h.c. (Poznan University of Economics)
Università degli Studi di Trieste
Dipartimento di Economia e Merceologia delle Risorse Naturali e della Produzione
Professore Ordinario
Via Valerio 6
I - 34100 Trieste
Tel: +39 40 6767018
Fax: +39 40 6763215

Private address
v. Raffaello Sanzio 36
I - 34128 Trieste
Tel: +39 40 54343

Fields of interest
Dairy, amino acids, lipids and fatty acids, HPLC, analytical chemistry, consumer research, quality, quality assurance, HACCP

CAMPANELLA, Luigi Full Professor
University of Rome
Department of Chemistry
Research Leader, Full Professor of Analytical Chemistry
P.le Aldo Moro 5
I - 00185 Roma
Tel: +39 6 49913744

Fax: +39 6 49913725

Private address
v. Cesare Abba 15
I - 00141 Roma
Tel: +39 6 8274268

Fields of interest
Oils and fats, water, amino acids, antioxidants, polyphenols, pesticides, electrochemistry, sensor technology, analytical chemistry, biotechnology

CHIAVARO, Emma Dr.
University of Parma
Department of Organic Chemistry and Industrial Chemistry
Researcher
Viale delle Scienze
I - 43100 Parma
Tel: +39 521 905406
Fax: +39 521 905472

Private address
S. Quarta 6
I - 43100 Parma
Tel: +39 521 244116

Fields of interest
Coffee, tea, cocoa, meat and meat products, poultry, spices, amino acids, mycotoxins, HPLC, analytical chemistry, high pressure technology, irradiation

CORRADINI, Claudio Dr.
C.N.R. Institute of Chromatography
Researcher
Area della Ricera di Roma, P.O. Box 10
I - 00016 Monterotondo Stazione (Roma)
Tel: +39 06 90672258
Fax: +39 06 90672269
E-Mail: corradcl@mlib.cnr.it
Private address
Via S. Giovanna Elisabetta 24
I - 00189 Roma
Tel: +39 06 3315466

Fields of interest
Beverages, bread and cereals, fruits and vegetables, sugar, honey, sugar alcohols, sweeteners, amino acids, antioxidants, carbohydrates, dietary fibres, fat replacers, Maillard reaction products, CE, HPLC, TLC, adulteration, analytical chemistry, food composition, quality

DAMIANI, Pietro Professor
Università degli Studi
Instituto di Chimica Bromatologica
Director and Full Professor of Food Chemistry
Via Romana

I - 06126 Perugia
Tel: +39 75 5853921
Fax: +39 75 31144
E-Mail: dapi@unipg.it
URL: dapi@unipg.it

Fields of interest
Fish and fish products, oils and fats, lipids and fatty acids, oxidation reactions, pesticides, chemometrics, GC, HPLC, MS, spectroscopy, TLC, adulteration, analytical chemistry, bioavailability, enzymology, food composition, quality, biotechnology

DI LUCCIA, Aldo Dr.
Instituto di Scienze dell' Alimentazione
Researcher
Via Roma 52 A-C
I - 83100 Avellino
Tel: +39 0825 781600
Fax: +39 0825 781585
E-Mail: adl@isa.av.cnr.it

Private address
Nicolo' Marcello Venuti 6
I - 80056 Errolano
Tel: +39 081 7393936
Fax: +39 081 7393936

Fields of interest
Dairy, meat and meat products, poultry, amino acids, lipids and fatty acids, Maillard reaction products, proteins, CE, electrophoresis, GC, HPLC, MS, spectroscopy, TLC, analytical chemistry, enzymology, food composition, quality, drying, preservation

DI NATALE, Corrado Dr.
University of Rome "Tor Vergata"
Department of Electronic Engineering
Researcher
Via di Tor Vergata
I - 00133 Roma
Tel: +39 6 72594408
Fax: +39 6 2020519
E-Mail: dinatale@eln.uniroma2.it

Fields of interest
Fish and fish products Fruits and vegetables, oils and fats, wine, chemometrics, sensor technology

DOSSENA, Arnaldo Associate Professor
University of Parma
Department of Organic Chemistry and Industrial Chemistry
Professor
Viale delle Scienze
I - 43100 Parma
Tel: +39 521 905555
Fax: +39 521 905472

Private address
Via Bertani 6
I - 43035 Felino (Parma)
Tel: +39 521 835532

Fields of interest
Dairy, meat and meat products, poultry, amino acids, antioxidants, mycotoxins, CE, HPLC, analytical chemistry, chemical reactions, quality

---

EVANGELISTI, Filippo Degree in Industrial Chemistry
University of Genova
Department of Food and Pharmaceutical Analysis and Technology
Professor
Via Brigata Salerno (Ponte)
I - 16147 Genova
Tel: +39 10 353 2603, 2647
Fax: +39 10 353 2684
E-Mail: evangelisti@dicfa.unige.it

Private address
Via Busalletta 19/1
I - 16010 Sant' Olcese (Genova)
Tel: +39 10 7092483

Fields of interest
Dairy, oils and fats, Maillard reaction products, oxidation reactions, polyphenols, steroids, GC, HPLC, food composition, quality

---

EVIDENTE, Antonio Professor
Università di Napoli "Frederico II"
Dipartimento die Scienze Chimico Agrarie
Professor of Organic Chemistry
Via Universitá 100
I - 80055 Portici
Tel: +39 081 7885224
Fax: +39 081 7755130

Private address
Via del Gran Paradiso18
I - 80144 Napoli
Tel: +39 81 5438283

Fields of interest
Bread and cereals, starch, amino acids, antimicrobials, aroma active compounds, biopolymers, carbohydrates, polyphenols, proteins, alkaloids, mycotoxins, electrophoresis, GC, HPLC, MS, spectroscopy, TLC, analytical chemistry, chemical reactions, food composition, microbiology, biotechnology

---

FAVRETTO, Luciano Degree in Chemistry
Università degli Studi di Trieste
Dipartimento di Economia e Merceologia delle Risorse Naturali e della Produzione
Full Professor of Commodity Science, Director of Department
Via Valerio 6
I - 34100 Trieste
Tel: +39 40 6767085
Fax: +39 40 6763215
E-Mail: luciano.favrello@econ.univ.trieste.it

Private address
dell'Eremo 74
I - 34139 Trieste
Tel: +39 40 943098

Fields of interest
Dairy, fish and fish products, fruits and vegetables, oils and fats, aroma active compounds, trace elements, AAS, chemometrics, food composition, quality

---

FERRARA, Lydia Prof. Associato
Università di Napoli
Dipartimento di Chimica Farmaceutica e Tossicologica
Docente
Via Domenico Montesano 49
I - 80131 Napoli
Tel: +39 81 7486611
Fax: +39 81 7486610

Private address
Via Michelangelo da Caravaggio 73
I - 80126 Napoli
Tel: +39 81 641571

Fields of interest
Cosmetics, fruits and vegetables, water, amino acids, lipids and fatty acids, alkaloids, HPLC, TLC, analytical chemistry, food composition

---

GABRIELLI FAVRETTO, Luciana Degree in Pharmacy
Università degli Studi di Trieste
Dipartimento di Economia e Merceologia delle Risorse Naturali e della Produzione
Full Professor in Food Chemistry
Via Valerio 6
I - 34100 Trieste
Tel: +39 40 6767025
Fax: +39 40 6763215
E-Mail: luciana.gabrielli@econ.univ.trieste.it

Private address
dell'Eremo 74
I - 34139 Trieste
Tel: +39 40 943098

Fields of interest
Dairy, fish and fish products, oils and fats, aroma active compounds, lipids and fatty acids, trace elements, AAS, chemometrics, GC, food composition

GATTI, Gian Carlo Dr.
NEOTRON S.R.L.
Director
Stradello Aggazzotti 104
I - 41010 S. Maria Di Mugano, Modena
Tel: +39 059 461711
Fax: +39 059 461777
E-Mail: neotron@neotron.it

Private address
Collina 210
I - 41058 Vignola (MO)
Tel: +39 59 774418

Fields of interest
Transgenic foods, additives, mycotoxins, PCBs, pesticides, sensory analysis, food composition, quality, packaging, MAP, quality assurance, HACCP

IORI, Renato Dr.
Instituto Sperimentale per le Colture Industriali
Senior Researcher
Via di Corticella 133
I - 40129 Bologna
Tel: +39 051 6316849
Fax: +39 051 374857
E-Mail: istsci@piperbole.bologna.it

Private address
Via Brenta 32
I - 41013 Castelfranco Emilia (Modena)
Tel: +39 059 924478

Fields of interest
Fruits and vegetables, antimicrobials, antioxidants, biopolymers, HPLC, Polarygraphy, spectroscopy, analytical chemistry, enzymology, biotechnology

LUNEIA, Roberto Ph.D.
University of Perugia, Institute of Food Chemistry
Research to Contract
Via San Constanzo, c/o Orto Botanico
I - 06126 Perugia
Tel: +39 075 5853917
Fax: +39 075 31144
E-Mail: roblu@cline.it

Private address
Via Foligno 2
I - 06058 San Terenziano (PG)
Tel: +39 0742 98942
Fax: +39 0742 98942

Fields of interest
Oils and fats, lipids and fatty acids, electrophoresis, GC, HPLC, MS, spectroscopy, TLC, adulteration, analytical chemistry, nutrition, quality

MARCHELLI, Rosangela Prof., Dr.
University of Parma
Department of Organic Chemistry and Industrial Chemistry
Dean of the Faculty of Agriculture
Parco Area delle Scienze n. 17/A
I - 43100 Parma
Tel: +39 521 905410
Fax: +39 521 905472
E-Mail: marchell@unipr.it

Private address
Borgo P. Cocconi 2
I - 43100 Parma
Tel: +39 521 237904

Fields of interest
Fruits and vegetables, meat and meat products, poultry, transgenic foods, amino acids, antioxidants, proteins, mycotoxins, CE, HPLC, quality

MARINI, Domenico Dr.
Dip. Dogane - Dir. Cent. Analisi Merceologica e Laboratorio Chimico
Head (Dirigente)
Via Carucci 71
I - 00143 Roma
Tel: +39 6 50957337
Fax: +39 6 50244114

Private address
Lungotevere Flaminio 44
I - 00196 Roma
Tel: +39 6 3224839

Fields of interest
Dairy, oils and fats, ethanol, lipids and fatty acids, chemometrics, HPLC, adulteration, food composition, quality assurance, HACCP

MARTELLI, Aldo Dr., Dean of Faculty of Pharmacy
University of Piemonte Orientale
Faculty of Pharmacy
Full Professor in Food Chemistry
Francesco Ferrucci
I - 28100 Novara
Tel: +39 0321 657610
Fax: +39 0321 657621
E-Mail: martelli@pharm.no.unipmn.it

Private address
Augusto Righi 16
I - 28100 Novara
Tel: +39 321 453617

Fields of interest
Coffee, tea, cocoa, dairy, transgenic foods, dietary fibres, polyphenols, biogenic amines, HPLC, PCR, biotechnology, quality assurance, HACCP

MAZZEI, Franco Dr.
University of Rome "La Sapienza"
Department of Pharmacology and General Physiology
Researcher
P.le Aldo Moro 5
I - 00185 Roma
Tel: +39 6 49913225
Fax: +39 6 49913888

Private address
Via E. Giovenale 42
I - 00176 Roma
Tel: +39 6 2757196
Fax: +39 6 2757196

Fields of interest
Wine, biogenic amines, marine toxins, pesticides, electrochemistry, sensor technology, analytical chemistry, toxicology

---

MERLINI, Lucio Degree Industrial Chemistry, Professor
Università di Milano
Dipartimento di Scienze Molecolari Agroalimentari
Professor
Via Celoria 2
I - 20133 Milano
Tel: +39 02 2663662
Fax: +39 02 70633062
E-Mail: lucio.merlini@unimi.it

Private address
Via C. Crivelli 14
I - 20122 Milano
Tel: +39 02 58322407

Fields of interest
Sweeteners, aroma active compounds

MORET, Ivo Professor
Università di Venezia
Dipartimento di Scienze Ambientali
Teaching and Research
Calle Larga S. Marta 2137
I - 30123 Venezia
Tel: +39 41 2578506
Fax: +39 41 2578549
E-Mail: moret@unive.it

Private address
Via Cima 18/L
I - 31020 San Pietro di Feletto

Fields of interest
Beverages, water, wine, aroma active compounds, PAHs, PCBs, pesticides, chemometrics, GC, analytical chemistry

---

PALLA, Gerardo Associate Professor
University of Parma
Department of Organic Chemistry
Food Chemistry
Viale delle Scienze
I - 43100 Parma
Tel: +39 521 905555
Fax: +39 521 905472
E-Mail: palla@unipr.it

Private address
Viale dei Mille 22
I - 42100 Reggio Emilia
Tel: +39 522 433446

PERTOLDI MARLETTA, Giuliana Degree in Pharmacy
Università di Trieste
Dipartimento di Economia e Merceologia delle Risorse Naturali e della Produzione
Full Professor in Food Chemistry
Via A. Valerio 6
I - 34100 Trieste
Tel: +39 40 6767089, 6767083
Fax: +39 40 6763215
E-Mail: giuliana.pertoldi@econ.univ.trieste.it

Private address
V. 10 Ospedale Militare 2/1
I - 34127 Trieste
Tel: +39 40 573887

Fields of interest
Dairy, fish and fish products, spices, trace elements, toxic trace elements, AAS, GC, food composition

---

PISCIOTTA, Gennaro Agricultural Expert
Istituto Professionale di Stato per L'Agricoltura e L' ambiente "Filippo Silvestri"
Teacher of Chemical Agricultural Laboratory
Via Domitiana 152
I - 80072 Licola Pozzuoli (Napoli)
Tel: +39 081 8678156
Fax: +39 081 8678156
E-Mail: genpis@tin.it

Private address
Via Lago Patria 104
I - 80014 Giugliano (Napoli)
Tel: +39 081 5091348

Fields of interest
Dairy, fruits and vegetables, legumes, oils and fats, water, wine, chemometrics, sensory analysis, spectroscopy, adulteration, analytical chemistry, chemical reactions, food composition, nutrition, quality, fermentation, quality assurance, HACCP, rheology

RESTANI, Patrizia Prof.
University Milano, Inst. Pharmacological Sciences
Teaching Food Chemistry, Researcher
Via Balzaretti 9
I - 20133 Milano
Tel: +39 02 20488225350
Fax: +39 02 20488260, 29404961
E-Mail: patrizia.restani@unimi.it

Private address
Corso XXII Marzo 43
I - 20129 Milano
Tel: +39 02 70121789

Fields of interest
Dairy, additives, proteins, mycotoxins, pesticides, GC, HPLC, MS, allergology, toxicology

SALVO, Francesco Professor
University of Messina
Department of Organic and Biological Chemistry
Professor of Food Chemistry
Salita Sperone 31, P.O. Box 46
I - 98166 St. Agata di Messina
Tel: +39 090 6765183
Fax: +39 090 393895
E-Mail: fsalvo@isengard.unime.it

Private address
Viale Regina Elena 97/C
I - 92121 Messina
Tel: +39 090 40504

Fields of interest
Fish and fish products, oils and fats, lipids and fatty acids, polyphenols, pesticides, GC, HPLC, food composition, quality, hygiene

SFORZA, Stefano Ph.D.
University of Parma
Department of Organic and Industrial Chemistry
Research Assistant
Parco Area delle Scienze n. 17/A
I - 43100 Parma
Tel: +39 0521 905555, 905406
Fax: +39 0521 905472
E-Mail: sforza@ipruniv.cce.unipr.it

Private address
B. Go al Collegio Maria Luigia 9
I - 43100 Parma
Tel: +39 0521 207628

Fields of interest
Dairy, meat and meat products, poultry, amino acids, biopolymers, proteins, GC, HPLC, MS, spectroscopy, chemical reactions, food composition

TATEO, Fernando Professor
Di. PRO. VE. Section DIFCA
Via Celoria 2
I - 20133 Milano
Tel: +39 02 26607227
Fax: +39 02 26607234
E-Mail: fernando.tateo@unimi.it

Private address
Via E. De Marchi 8
I - 20125 Milano
Tel: +39 2 66986714

Fields of interest
Spices, sweeteners, aroma active compounds, Maillard reaction products, GC, HPLC, MS, sensory analysis, food composition

UCCELLA, Nicola Antonio MA, Ph.D., C.Chem. FRSC
University of Calabria
Professor of Organic Chemistry
Via P. Bucci
I - 87030 Rende
Tel: +39 984 492049
Fax: +39 984 492116
E-Mail: m.uccella@unical.it

Private address
Via Po 44
I - 87030 Rende

Fields of interest
Cosmetics, oils and fats, antioxidants, lipids and fatty acids, polyphenols, HPLC, MS, sensory analysis, sensor technology, food composition, quality

VALENTINI, Giuseppa Associate Professor
Department of Chemical Sciences
Via S. Agostino 1
I - 63032 Camerino (MC)
Tel: +39 737 402254
Fax: +39 737 637345
E-Mail: valentin@camserv.unicam.it

Private address
Caselunghe 5
I - 62032 Camerino (MS)
Tel: +39 339 4786361

Fields of interest
Aroma active essential oils, drugs, GC, MS, hydrodistillation

ZIINO, Marisa Degree in Biological Siences
University of Messina
Department of Organic and Biological Chemistry
Professor of Food Chemistry
Salita Sperone 31, P.O. Box 46
I - 98166 St. Agata di Messina

Tel: +39 90 6765167
Fax: +39 90 392840

Private address
A. Salandra 22
I - 98100 Messina
Tel: +39 90 694201

Fields of interest
Dairy, fish and fish products, meat and meat products, poultry, oils and fats, lipids and fatty acids, proteins, GC, HPLC, MS, TLC, food composition

---

ZUNIN, Paola Degree in Pharmaceutical Chemistry
University of Genova
Department of Food and Pharmaceutical Analysis and Technology
Researcher
Via Brigata Salerno (Ponte)
I - 16147 Genova
Tel: +39 10 353 2603, 2646
Fax: +39 10 353 2684
E-Mail: zunin@dictfa.unige.it

Private address
Via delle Ginestre 35/3
I - 16137 Genova
Tel: +39 10 885891

Fields of interest
Dairy, oils and fats, Maillard reaction products, oxidation reactions, polyphenols, steroids, GC, HPLC, food composition, quality

Liechtenstein

MATT, Helmuth Dr., Dipl.Chem., Eidg. Dipl.
Lebensmittelchemiker
Labor Dr. H. Matt AG
Inhaber des Labors
Im alten Riet 36, Postfach 503
FL - 9494 Schaan
Tel: +423 2333833
Fax: +423 2333835
E-Mail: dr.matt@labor.lol.li

Private address
FL - 9494 Schaan

Lithauania

VENSKUTONIS, Petras Rimantas PhD, Dipl. Eng.
Kaunas University of Technology
Department of Food Technology
Associate Professor, Head of Department
Radvilenu pl. 19
LT - 3028 Kaunas
Tel: +370 7 456426
Fax: +370 7 456647
E-Mail: rimas.venskutonis@ctf.ktu.lt

Private address
Topoliu 2-17
LT-3031 Kaunas
Tel: +370 7 792636

Fields of interest
Fruits and vegetables, oils and fats, spices, additives, aroma active compounds, GC, food composition, preservation, quality assurance, HACCP

Netherlands

AALBERSBERG, Willem Y. Dr.
Aalbersberg Consult BV
Director
Denmenlaan 19
NL - 6705 BX Wageningen
Tel: +31 317 420028
Fax: +31 317 413720
E-Mail: aalbersbergconsult@captiolonline.nl

Fields of interest
Dairy, fish and fish products, meat and meat products, poultry, potatoes, transgenic food, proteins, biotechnology, fermentation, minimal processing, thermal processing

---

BAARS, Aalbert Jan Dr.
Centre for Substances and Risk Assessment (CSR)
National Institute of Public Health and the Environment (RIVM)
Senior Toxicologist
P.O. Box 1
NL - 3720 - BA Bilthoven
Tel: +31 30 2743689
Fax: +31 30 2744401
E-Mail: aj.baars@rivm.nl

Private address
Kampstraat 18
NL - 6711 BS Ede
Tel: +31 318 616024
Fax: a.j.baars@worldonline.nl

Fields of interest
Marine toxins, mycotoxins, PAHs, PCBs, pesticides, plant toxins, toxic trace elements, toxicology, risk assessment

Netherlands

BEENTJES, Pieter Drs.
Foodconsult
Director
Ferd Bolstraat 9
NL - 7901 CE Hoogexeen
Tel: +31 52 8062372
Fax: +31 52 8062372

Private address
Ferd Bolstraat 9
NL - 7901 CE Hoogexeen
Tel: +31 52 8262372
Fax: +31 52 8062372

BELJAARS, Paul Ph.D., M.Sc.
Regional Inspectorate W&V South
Research & Development, Head
Rijzertlaan 19
NL - 5223 JS Den Bosch
Tel: +31 40 2911500
Fax: +31 40 2911600
E-Mail: paul.beljaars@inspectwv.nl

Private address
Kampseveld 11
NL - CB Den Dungen
Tel: +31 73 5943004
Fax: +31 73 5943004

Fields of interest
Beer, beverages, bread and cereals, coffee, tea, cocoa, oils and fats, spices, spirits, starch, sugar, honey, sugar alcohols, transgenic food, wine, additives, amino acids, antioxidants, carbohydrates, dietary fibres, lipids and fatty acids, proteins, trace elements, vitamins, alkaloids, biogenic amines, heterocyclic aromatic amines, hormones, mycotoxins, AAS, electrophoresis, GC, HPLC, MS, PCR, Polarography, spectroscopy, TLC, adulteration, analytical chemistry, consumer research, food composition, management, microbiology, nutrition, toxicology, biotechnology, genetic engineering, preservation, quality assurance, HACCP

BONTENBAL, Edwin M.Sc., MBA
PURAC
Marketing Manager
Achelsedyk 46
NL - 4700 AA Gorinchem
Tel: +31 183 695882
E-Mail: e.w.bontenbal@purac.com

Private address
Nudestraat 6
NL - Wageningen
Tel: +31 7 319325

Fields of interest
Meat and meat products, poultry, additives, antimicrobials, minerals, bioavailability, chemical reactions, management, microbiology, nutrition

CLARK, David B.Sc., Ph.D.
DMV International
Manager, Research and Analysis Department
NLB Laan 80
NL - 5460 BA Veghel
Tel: +31 413 372310
Fax: +31 413 367162
E-Mail: clarkd@dmv-international.com

Fields of interest
Beer, beverages, dairy, amino acids, antimicrobials, proteins, electrophoresis, GC, HPLC, spectroscopy, bioavailability, enzymology, nutrition, fermentation, rheology, emulsions, foams

CORNELESE, Johan Ing.
Ministry of Agriculture
Department for Science and Knowledge Transfer
Senior Officer for Food, Non-Food and Fisheries Research
P.O. Box 20401
NL - 2500 EK Den Haag
Tel: +31 70 3785597
Fax: +31 70 3786183
E-Mail: j.a.cornlese@dwk.agro.nl

Fields of interest
Management

DE JONG, Jacob Dr.
State Institute for Quality Control of Agricultural Products (RIKILT-DLO)
Project Manager Official Control
Bornsesteeg 45
NL - 6708 PD Wageningen
Tel: +31 317 475400
Fax: +31 317 417717

Private address
Kotehoven 11
NL - 6721 SX Bennekom
Tel: +31 318 418458

Fields of interest
Dairy, meat and meat products, poultry, oils and fats, antimicrobials, drugs, hormones, HPLC, adulteration, analytical chemistry, regulative issues

DE KRUIF, Kees C.G. Prof., Dr.
NIZO Food Research
Head of Department, Leader of Programme
Kernhemseweg 2, P.O. Box 20
NL - 6710 BA Ede
Tel: +31 318 659511
Fax: +31 318 650400
E-Mail: dekruif@nizo.nl

Private address
Parklaan 28; NL - 3931 KK Woudenberg
Tel: +31 33 2862193

Fields of interest
Dairy, biopolymers, proteins, thermal anlaysis, management, fouling and cleaning, high pressure technology, rheology

DE RUITER, Gerhard A. Dr.
NIZO Food Research
Groupleader Ingredient Technology
P.O. Box 20
NL - 6710 BA Ede
Tel: +31 318 659511
Fax: +31 3318 650400
E-Mail: deruiter@nizo.nl

Fields of interest
Beverages, dairy, oils and fats, carbohydrates, fat replacers, proteins, HPLC, analytical chemistry, enzymology, rheology

ELLEN, Geert Dr.
NIZO Food Research
Head of Department of Analytical Chemistry
Kernhemseweg 2, P.O. Box 20
NL - 6710 BA Ede
Tel: +31 318 659511
Fax: +31 318 650400
E-Mail: ellen@nizo.nl

Private address
P.C. Hooflaan 28
NL - 3750 AH Zeist
Tel: +31 30 6951707

Fields of interest
Dairy, antimicrobials, lipids and fatty acids, minerals, trace elements, toxic trace elements, AAS, GC, Polarography, analytical chemistry, quality, toxicology, quality assurance, HACCP

ESHUIS, Dolf F. Ir.
Food & Technology Consult GKVC
Consultant
Piet Heynlaan 42
NL - 4819 AJ Breda
Tel: +31 76 5217699
Fax: +31 76 5144571

Private address
Piet Heynlaan 42
NL - 4819 AJ Breda
Tel: +31 76 5217699
Fax: +31 76 5144571

Fields of interest
Fruits and vegetables, legumes, potatoes, additives, antioxidants, oxidation reactions, polyphenols, minimal processing, preservation, food legislation

GROOTHUIS, Dirk G. D.V.M., Ph.D.
General Inspectorate for Health Protection
Senior Officer of Public Health, Veterinarian
P.O. Box 16.108
NL - 2500 BC Den Haag
Tel: +31 70 3406927
Fax: +31 70 3405435
E-Mail: dirk.groothuis@inspectwv.nl

Fields of interest
Eggs, fish and fish products, meat and meat products, poultry, antimicrobials, marine toxins, microbiology, quality assurance, HACCP

HAVENAAR, Robert D.V.M., Ph.D.
TNO Nutrition and Food Research Institute
Product Manager GI Research
Utrechtseweg 48, P.O. Box 360
NL - 3700 AJ Zeist
Tel: +31 30 6944726
Fax: +31 30 6944928
E-Mail: havenaar@voeding.tno.nl

Private address
Onde Arnhemseweg 225
NL - 3705 BC Zeist
Tel: +31 30 6961471

Fields of interest
Carbohydrates, minerals, proteins, vitamins, bioavailability, metabolism, microbiology, nutrition, quality, GI tract models

HERMUS, Rudolph J.J. Prof., Dr. Ir.
TNO Nutrition and Food Research Institute
P.O. Box 360/Utrechtseweg 48
NL - 3700 AJ Zeist
Tel: +31 30 6944763
Fax: +31 30 6957952
E-Mail: hermus@stp.tno.nl

Private address
Waterstraat 3
NL - 6658 AA Beneden Leeuwen
Tel: +31 487 591485
Fax: +31 487 594385

Fields of interest
Transgenic food, antioxidants, carotenoids, lipids and fatty acids, vitamins, plant toxins, bioavailability, food composition, nutrition, toxicology

KAN, Kees (C.A.) Drs.
Institute for Animal Science and Health
Senior Scientist

Edelhertweg 15
NL - 8200 AB Lelystad
Tel: +31 320 238238
Fax: +31 320 238050
E-Mail: c.a.kan@id.wag.uz.nl

Private address
Sneenwklokjesveld 15A
NL - 8255 JT Swifterbant
Tel: +31 321 323550
Fax: +31 321 324914

Fields of interest
Eggs, meat and meat products, poultry, antimicrobials, drugs, mycotoxins, PCBs, pesticides

KERKVLIET, Jacob Drs.
Inspectorate for Health Protection
Hoogte Kadyk 401
NL - 1018 BK Amsterdam
Tel: +31 20 5244600
Fax: +31 20 5244700
E-Mail: jaap.kerkvliet@inspectwv.nl

Private address
Rustenburgherweg 16
NL - 2061 JB Bloemendaal
Tel: +31 23 5257314

Fields of interest
Fruits and vegetables, legumes, potatoes, spices, sugar, honey, sugar alcohols, microscopy, adulteration, food composition, nutrition, toxicology

KLAARENBEEK, Tineke Ir.
Pepsilo Foods International Europe
Frito Lay Europe
Regulatory & Nutrition Manager
Zonnebaan 35
NL - 3600 BA Maarssen
Tel: +31 30 2473811
Fax: +31 30 2473651
E-Mail: tineken.klaarenbeek@intl.fritolay.com

Private address
Vechtensteinlaan 21
NL - 3601 CM Maarssen
Tel: +31 34 6569612

Fields of interest
Oils and fats, potatoes, sensory analysis, nutrition, regulative issues, quality assurance, HACCP

LANGENDAM, Johannes Drs. Chemistry
Arnhem-Nijmegen University of Professional Education
Director Department of Nutrition
P.O. Box 6960

NL - 6503 GL Nijmegen
Tel: +31 24 3277888
Fax: +31 24 3600863
E-Mail: hans.langendam@mo.han.nl

Private address
Burg. v. d. Berghstr. 45
NL - 6512 DD Nümegen
Tel: +31 24 3230287

Fields of interest
Management, nutrition, education

LAUWAARS, Margreet Chem.cand.RUL
AOAC International
Liaison coordinator
P.O. Box 153
NL - 6720 AD Bennekom
Tel: +31 318 418725
Fax: +31318 418759
E-Mail: lauwaars@worldonline.nl

Private address
Grietjeshof 69
NL - 6721 UH Bennekom
Tel: +31 318 418725
Fax: +31 318 418359

Fields of interest
Scientific associations

LELOUX, Mirjam Dr.
RIKILT
Deputy Director
Bornsetsteeg 45, P.O. Box 230
NL - 6700 AE Wageningen
Tel: +31 317 475400
Fax: +31 317 417717
E-Mail: m.s.leloux@rikilt.wag-ur.nl

Private address
Bezuidenhout 14
NL - 6711 J2 Ede
Tel: +31 318 410922

Fields of interest
Food chemistry, research management

LOSSONCZY VON LOSONCZ, Thomas M.Sc.
Haagse Hogeschool
Senior Teacher
NL - 2521 EN Den Haag
Tel: +31 70 4458289
Fax: +31 70 4458125
E-Mail: lov@sem.hhs.nl

Private address
Van Beinumstraat 363
NL - 2551 HW Den Haag

Tel: +31 70 3910798

Fields of interest
Oils and fats, lipids and fatty acids, vitamins, mycotoxins, nutrition, quality, biotechnology, quality assurance, HACCP

MACLEAN, Wim Ir., Drs.
Borculo Whey Products
Manager Quality Assurance
Hanzeplein 25
NL - 8017 JD Zwolle
Tel: +31 38 4677444
Fax: +31 38 4677555

Private address
De Hofvoogd 11
NL - 7271ZP Borculo
Tel: +31 545 274761

Fields of interest
Dairy, quality, quality assurance, HACCP

NORTHOLT, Martin Dr.
TNO Nutrition and Food Research Institute
Productmanager Microbiological Food Safety
P.O. Box 360/Utrechtseweg 48
NL - 3700 AJ Zeist
Tel: +31 30 6944632
Fax: +31 30 6944901
E-Mail: northolt@voeding.tno.nl

Private address
Leinweberlaan 25
NL - 3971 KX Driebergen
Tel: +31 343 513098

Fields of interest
Dairy, water, antimicrobials, biogenic amines, mycotoxins, microbiology, fermentation, hygiene, minimal processing, preservation, quality assurance HACCP, storage, thermal processing

NOTEBORN, Hubert Dr.
RIKILT Wageningen - UR
Senior Scientist, Food Toxicology
Bornsesteeg 45
NL - 6708 PD Wageningen
Tel: +31 317 475400, 475462
Fax: +31 317 417717
E-Mail: h.p.j.m.noteborn@rikilt.wag-ur.nl

Private address
Atalantaberm 34
NL - WB Houten
Tel: +31 30 6373330
Fax: +31 30 6373330

Fields of interest
Fruits and vegetables, transgenic foods, polyphenols, proteins, alkaloids, hormones, plant toxins, chemometrics, HPLC, bioavailability, food composition, toxicology, biotechnology, genetic engineering

OLIEMAN, Cornelis Dr., Ir.
Netherlands Institute for Dairy Research (NIZO)
Scientist in Analytical Chemistry
Kernhemseweg 2, P.O. Box 20
NL - 6710 BA Ede
Tel: +31 318 659511
Fax: +31 318 650400
E-Mail: olieman@nizo.nl

Private address
Roolvinkpark 1
NL - 6716 EG Ede
Tel: +31 318 624927

Fields of interest
Dairy, amino acids, carbohydrates, proteins, CE, HPLC, adulteration, food composition

ROEL, Peter M.Sc.
Int. Flavors & Fragrances Netherlands
Application MNG. Sweet & Beverages Lab. EAME
Liebergerweg 72-98
NL - 1221 JT Hilversom
Tel: +31 35 883911
Fax: +31 35 883202
E-Mail: peter.roel@iff.com

Private address
Bachlaan 9
NL - 1411 JB Naarden
Tel: +31 35 6947405

Fields of interest
Beverages, fruits and vegetables, additives, aroma active compounds, management, biotechnology, fermentation, preservation

ROMBOUTS, Frank Prof, Dr., IR.
Wageningen Agricultural University
Wageningen University
Professor in Food Microbiology
P.O. Box 8129
NL - 6700 ER Wageningen
Tel: +31 317 482233
Fax: +31 317 484893

Private address
Polderstraat 6
NL - 6666 LD Heteren
Tel: +31 26 4722593
Fax: +31 26 4723914

Fields of interest
Antimicrobials, carbohydrates, mycotoxins, microbiology, fermentation, hygiene, minimal processing, preservation, quality assurance, HACCP

ROOZEN, Jacques Dr. Ir.
Wageningen Agricultural University
Department of Agrotechnology and Food Science
Associate Professor
Bomenweg 2
NL - 6703 HD Wageningen
Tel: +31 317 482234
Fax: +31 317 484893
E-Mail: jacques.roozen@chem.fdsci.wag-ur.nl

Private address
Abersonlaan 1
NL - 6703 GD Wageningen
Tel: +31 317 410625

Fields of interest
Beverages, sweeteners, antioxidants, aroma active compounds, GC, sensory analysis, chemical reactions

ROPS, Wichard Drs.
Delta College
ROC Zeeland
Director Horeca
Merwedestraat 1
NL - 4335 XR Middelburg
Tel: +31 11 8623951
Fax: +31 11 8626556

Private address
Diepenbrocklaan 27
NL - 4702 HE Roosendaal
Tel: +31 16 5558140
Fax: +31 16 5558140

Fields of interest
Beer, coffee, tea, cocoa, wine, fat replacers, PCBs, sensory analysis, food composition, microbiology, biotechnology, hygiene

RUITER, Adriaan Dr.
Scientifix Consultancy
Emeritus Professor of Food Chemistry
E-Mail: a.ruiter@hetnet.nl

Private address
Willem de Zwijgerlaan 5
NL - 3722 JR Bilthoven
Tel: +31 30 2287164

Fields of interest
Fish and fish products, drugs, marine toxins, PCBs, food composition, nutrition

SCHUTTE, Leonard Ph.D., Dr.
Quest International
Unit Head
P.O. Box 2
NL - 1400 CA Bussum
Tel: +31 35 6992817
Fax: +31 35 6953321
E-Mail: leonard.schutte@questintl.com

Private address
Teding van Berkhoutlaan 47
NL - 2111 ZB Aerdenhout
Tel: +31 23 5440114
Fax: +31 23 5441047

Fields of interest
Dairy, oils and fats, aroma active compounds

SIEZEN, Roland J. Ph.D.
Netherlands Institute for Dairy Research (NIZO)
Research Manager Live Sciences
Kernhemseweg 2, P.O. Box 20
NL - 6710 BA Ede
Tel: +31 318 659 511, 542
Fax: +31 318 650400
E-Mail: siezen@nizo.nl

Private address
Eikelakkers 2
NL - 6711TE Ede
Tel: +31 318 619059

Fields of interest
Dairy, transgenic foods, antimicrobials, proteins, Peptides, enzymology, metabolism, biotechnology, genetic engineering, bioinformatics

STARK, Jacques Drs.
DSM Food Specialties
Head of Department and Fermentation
P.O. Box 1
NL - 2600 MA Delft
Tel: +31 15 2792226
Fax: +31 15 2792490
E-Mail: jacques.stark@dsm-group.com

Private address
Joost van Ospad 31
NL - 3069 ZM Rotterdam
Tel: +31 10 4214958

Fields of interest
Dairy, meat and meat products, poultry, antimicrobials, drugs, microbiology, biotechnology, preservation

STEINBUCH, Erwin HGRIC
Spenger Institute
Head Technology Department

Plantsoen 134
NL - 6701 AT Wageningen
Tel: 31 317 415767

Private address
Plantsoen 134
NL - 6701 AT Wageningen
Tel: +31 317 415767

Fields of interest
Fruits and vegetables, quality, biotechnology, drying, fermentation, freezing, minimal processing, preservation, quality assurance, HACCP, thermal processing

---

TIMMERMANS, Eric Ir.
Borculo Domo Ingredients
Manager R & D
P.O. Box 46
NL - 7270 AA Borculo
Tel: +31 545 256789
Fax: +31 545 256625
E-Mail: hjar.timmermans@bdi.nl

Private address
De Hofvoogal 7
NL - 7271 KE Borculo
Tel: +31 545 273374

Fields of interest
Dairy, carbohydrates, minerals, proteins, HPLC, sensory analysis, thermal anlaysis, enzymology, nutrition, drying, filtration

---

UIJTTENBOOGAART, Theo Agr.Eng. (Ir.)
ID-DLO
Institute for Animal Science and Health
Edelhertweg 15
NL - 8219 PH Lelystad
Tel: +31 320 238238
Fax: +31 320 238984
E-Mail: t.g.uijttenboogaart@id.wag-nr.nl

Private address
Dorpstraat 106
NL - 7361 AZ Beekbergen
Tel: +31 55 5062224
Fax: +31 55 5061193

Fields of interest
Eggs, meat and meat products, poultry, proteins, sensory analysis, thermal anlaysis, food composition, metabolism, quality, freezing, hygiene, packaging, MAP, preservation, storage

---

VAN BOEKEL, Tiny Dr.
Wageningen Agricultural University
Department of Food Science
Senior Lecturer/Associate Professor
Bomenweg 2
NL - 6703 HD Wageningen
Tel: +31 317 484281
Fax: +31 317 483669

Private address
Lawickse Allee 32
NL - 6707 AH Wageningen
Tel: +31 317 423158

Fields of interest
Dairy, fruits and vegetables, amino acids, antioxidants, biopolymers, Maillard reaction products, proteins, HPLC, chemical reactions, thermal processing

---

VAN DEN BOSCH, Gerrit Dr.
Cehave NV
Research-Director
N.C.B.-Laan 52
NL - 5460 BC Veghel
Tel: +31 41 3082345
Fax: +31 41 3082833
E-Mail: ccl@ccl.nl

Private address
Dommeloord 23
NL - 5283 LL Boxtel
Tel: +31 41 1674685

Fields of interest
Eggs, meat and meat products, poultry, oils and fats, amino acids, Animicrobials, antioxidants, aroma active compounds, carbohydrates, carotenoids, lipids and fatty acids, proteins, trace elements, vitamins, biogenic amines, mycotoxins, pesticides, toxic trace elements, AAS, GC, HPLC, sensory analysis, bioavailability, food composition, nutrition, biotechnology, fermentation, hygiene, quality assurance, HACCP, thermal processing

---

VAN DEN BROEK, AD Drs.
Sentron Europe BV
International Marketing & Sales Manager
Aan de Vaart 3
NL - 9301 ZH Roden
Tel: +31 50 13800
Fax: +31 50 16834
E-Mail: sentron@sentron.nl

Fields of interest
Sensor technology, analytical chemistry, ISFET technology for glass-free unbreakable pH-sensors

---

VAN DER SCHEE, Henk A. Dr.
Inspectorate for Health Protection
Head of Chemistry Department
Hoogtekadijk 401

Netherlands

NL - 1010BK Amsterdam
Tel: +31 20 5244600
Fax: +31 20 5244700
E-Mail: henk.van.der@inspectwv.nl

Fields of interest
Fruits and vegetables, pesticides, toxic trace elements, GC, HPLC, analytical chemistry

---

VAN DOKKUM, Wim Ph.D., Professor
TNO Nutrition and Food Research Institute
Freelance advisor
Nyenheim 11-17
NL - 3704 AC Zeist
Tel: +31 30 6944144
Fax: +31 30 6957952
E-Mail: w.dokkum@wxs.nl

Private address
Nyenheim 1117
NL - 3704 AC Zeist
Tel: +31 30 6992860
Fax: +3130 6992861

Fields of interest
Dietary fibres, minerals, trace elements, bioavailability, nutrition, functional foods

---

VAN POPPEL, Geert A.F.C. M.Sc., Ph.D.
TNO Nutrition and Food Research
Business Development Units BDU Food and Health
Senior Scientist
P.O. Box 360
NL - 3700 AJ Zeist
Tel: +31 30 6944761
Fax: +31 30 6959792

VAN RHYN, Hans Drs.
State Institute for Quality Control of Agricultural Products (RIKILT-DLO)
Senior Scientist, Residue Analysis
Bornsesteeg 45
NL - 6708 PD Wageningen
Tel: +31 317 475597
Fax: +31 317 417717
E-Mail: j.a.vanrhyn@rikilt.dlo.nl

Private address
Zwanebloemstraat 8
NL - 6871 WN Renkum
Tel: +31 317 317442

Fields of interest
Dairy, eggs, fish and fish products, meat and meat products, poultry, sugar, honey, sugar alcohols, antimicrobials, steroids, drugs, hormones, HPLC, MS, analytical chemistry, quality, regulative issues, quality assurance HACCP, food safety

---

VORAGEN, Fons Dr.
Wageningen Agricultural University
Laboratory of Food Chemistry
Professor of Food Chemistry
P.O. Box 8129
NL - 6700 EV Wageningen
Tel: +31 317 483209, 482888
Fax: +31 317 484893
E-Mail: fons.voragen@chem.fdsci.wau.nl

Private address
Sparrenbos 37
NL - 6705 BB Wageningen
Tel: +31 317 416976

Fields of interest
Bread and cereals, fruits and vegetables, biopolymers, carbohydrates, proteins, bioavailability, chemical reactions, enzymology, biotechnology, genetic engineering

---

WIJNGAARDS, Gerrit Dr.
TNO Nutrition and Food Research Institute
Senior Advisor Food Ingredients
Utrechtseweg 48
NL - 3704 HE Zeist
Tel: +31 30 6944144, 6944169
Fax: +31 30 6957224
E-Mail: wijngaards@voeding.tno.nl

Private address
Van Bijnenlaan 10
NL - 3972 GS Driebergen
Tel: +31 343 517543

Fields of interest
Meat and meat products, poultry, biopolymers, proteins, electrophoresis, spectroscopy, enzymology, food composition, biotechnology, rheology

---

ZAGT, Robert Drs.
Food Industrial Research Assistance Consult, Owner, freelance
Joh.Wagenaarlaan 21
NL - 2102 GA Heemstede
Tel: +31 23 5286316
Fax: +31 23 5294191

Private address
John. Wagenaarlaan 21
NL - 2102 GA Heemstede
Tel: +31 23 5286316
Fax: +31 23 5294191

Fields of interest
Dairy, bioavailability, microbiology, nutrition, hygiene, quality assurance, HACCP

Norway

BRATHEN, Gudmund
Norwegian Inst. for Food and Environmental Analysis
Research Project Manager
Nils Hansen V:4, P.O.Box 6166 Etterstad 0602
N - 0602 Oslo
Tel: +47 23 050500
Fax: +47 23 050501
E-Mail: gudmundbraathen@matanalyse.no

Private address
Singers V. 12
N - 1450 Nesoddtangen
Tel: +47 6 910695

Fields of interest
Dairy, fish and fish products, meat and meat products, poultry, oils and fats, water, lipids and fatty acids, proteins, vitamins, AAS, chemometrics, GC, HPLC, sensory analysis, spectroscopy, microbiology, quality

EKLUND, Trygve MBA, Ph.D., MSc, BSc
National Institute of Occupational Health
Director General
PO Box 8149 DEP
N - 0033 Oslo
Tel: +47 2319 5100
Fax: +47 2319 5201
E-Mail: trygve.eklund@stami.no

Private address
N - 1472 Fjellhamar
Tel: +47 6779 5084

Fields of interest
Antimicrobials, management, microbiology, hygiene, preservation

NORTVEDT, Ragnar Dr.Scient.
Directorate of Fisheries
Institute of Nutrition
Senior Researcher
P.O. Box 185, Sentrum
N - 5804 Bergen
Tel: +47 55 238252
Fax: +47 55 238095
E-Mail: ragnar.nortvedt@nutr.fiskeridir.no

Fields of interest
Fish and fish products, amino acids, lipids and fatty acids, minerals, vitamins, chemometrics, food composition, quality, smoking

SARPEID, Hans-Jacob Dr.scient.
Matforsk, Norwegian Food Research Institute
Research Scientist
Osloveien 1
N - 1430 Aas
Tel: +47 64 970313
Fax: +47 64 970333
E-Mail: hans-jacob.skarpeid@matforsk.no

Private address
Gamleveien 13
N - 1430 Aas
Tel: +47 64 944524

Fields of interest
Dairy, fish and fish products, meat and meat products, poultry, proteins, chemometrics, electrophoresis, adulteration, allergology, enzymology

SLINDE, Erik Prof.Dr.philos.
Institute of Marine Research
Research Director
N - 5817 Bergen
Tel: +47 55 238500
Fax: +47 55 236379
E-Mail: erik@imr.no

Private address
N - 5098 Bergen
Tel: +47 55 288852

SORHAUG, Terje Ph.D.
University of Norway
Department of Food Science
Professor
P.O. Box 5036
N - 1432 As
Tel: +47 64 948570
Fax: +47 64 946789
E-Mail: terje.sorhaug@inf.nlh.no

Private address
Hellinga 14B
N - 1430 As
Tel: +47 64 942959

Fields of interest
Dairy, proteins, enzymology, microbiology, biotechnology, fermentation

Poland

AMAROWICZ, Ryszard Ph.D.
Department of Food Chemistry, Division of Food Science, Institute of Animal Reproduction and Food Research of Polish Academy of Sciences
Assistant Professor
Tuwima 10, P.O. Box 55
PL - 10718 Olsztyn
Tel: +48 89 5234675

Fax: +48 89 5240124
E-Mail: amaro@food.pan.olsztyn.pl

Private address
Boenigka 13M8
PL - 10686 Olsztyn
Tel: +48 89 5413769

Fields of interest
Coffee, tea, cocoa, legumes, antioxidants, carbohydrates, polyphenols, proteins, HPLC, spectroscopy, TLC, chemical reactions

---

BACHMAN, Stefania Ph.D., D.Sc., Professor
Institute of Applied Radiation Chemistry
Professor
Wroblewski 15
PL - 93590 Lodz
Tel: +48 42 6313158
Fax: +48 42 6360245
E-Mail: mitr@mitr.p.lodz.pl

Private address
Al. Kosciuszki 56/2
PL-90428 Lodz
Tel: +48 42 6366258

Fields of interest
Potatoes, spices, starch, additives, carbohydrates, Maillard Reaction, mycotoxins, electrophoresis, irradiation, preservation

---

BYKOWSKI, Piotr Jan Ph.D., D.Sc.
Sea Fisheries Institute
Head of Fish Processing Department, Professor
Kollataja 1
PL - 81332 Gdynia
Tel: +48 58 205211
Fax: +48 58 202831
E-Mail: ztimp@mir.go
ztimp@mir.gdynia.pl

Private address
Piastowska 66b3
PL - 80363 Gdansk
Tel: +48 58 571965

Fields of interest
Fish and fish products, biogenic amines, PAHs, PCBs, pesticides, toxic trace elements, AAS, GC, HPLC, MS, freezing, quality assurance HACCP, dioxins

---

CHRZANOWSKA, Jozefa Ph.D., D.Sc.
Agricultural University of Wroclaw
Department of Animal Products Technology
Dean of Faculty of Food Technology
C.K. Norwida 25
PL - 50375 Wroclaw

Tel: +48 71 3205499
Fax: +48 71 3284124
E-Mail: jch@ozi.ar.wroc.pl

Private address
Canaletta 11
PL - 51650 Wroclaw
Tel: +48 71 3451766
Fax: +48 71 3489416

Fields of interest
Dairy, enzymology

CZARNECKA, Maria Dr.
University of Agriculture
Department of Food Technology
Assistant Professor
Wojska Polskiego 31
PL - 60624 Poznan
Tel: +48 61 8487282
Fax: +48 61 8487314
E-Mail: marcz@au.poznan.pl

Private address
Piatkowska 55/36
PL - 60648 Poznan
Tel: +48 61 8239954

Fields of interest
Legumes, potatoes, spirits, starch, sugar, honey, sugar alcohols, fermentation, preservation, storage

---

CZARNECKI, Zbgniew Dr.habil.
University of Agriculture
Department of Food Technology
Associate Professor
Wojska Polskiego 31
PL - 60624 Poznan
Tel: +48 61 8487285
Fax: +48 61 8487314
E-Mail: zbyczar@au.poznan,pl

Private address
Piatkowska 55/36
PL - 60648 Poznan
Tel: +48 61 8239954

Fields of interest
Beer, beverages, legumes, potatoes, spirits, starch, sugar, honey, sugar alcohols, ethanol, fermentation

---

DZIUBA, Jerzy Prof., Dr. habil.
Warmia and Masurian University in Olsztyn
Chair of Food Biochemistry
Deputy Dean for Research, Head of Chair
Kortowo, Plac Cieszynski 1
PL - 10957 Olsztyn
Tel: +48 89 5233715
Fax: +48 89 5240408

Private address
Iwaszkiewicza 16/22
PL - 10089 Olsztyn
Tel: +48 89 5276282

Fields of interest
Dairy, biopolymers, proteins, biogenic amines, electrophoresis, HPLC, spectroscopy, analytical chemistry, chemical reactions, enzymology, rheology

DZWOLAK, Waldemar Ph.D., M.Sc.
University of Agriculture and Technology
Department of Dairy Technology
Assistant Professor
Oczapowskiego 7
PL - 10957 Olsztyn
Tel: +48 89 5234472
Fax: +48 89 5233402
E-Mail: waldekdz@moskit.uwm.edu.pl

Private address
Zolniernska 14C/203
PL - 10561 Olsztyn
Tel: +48 89 5340796

Fields of interest
Dairy, amino acids, proteins, electrophoresis, quality, quality assurance, HACCP

GARNCAREK, Barbara Dr.
Academy of Economics
Food Biotechnology Department
Adjunct
Komandorska 118/120
PL - 53345 Wroclaw
Tel: +48 71 3680256
Fax: +48 71 3672778
E-Mail: zbgarnc@credit.ae.wroc.pl

Private address
Ciepla 16/12
PL - 50524 Wroclaw
Tel: +48 71 3733632

Fields of interest
Fruits and vegetables, starch, sweeteners, biopolymers, carbohydrates, HPLC, sensory analysis, food composition, microbiology, rheology

GOLACHOWSKI, Antoni Ph.D.
Agricultural University of Wroclaw
Department of Food Storage and Technology
Assistant Professor
Norwida 25
PL - 50375 Wroclaw
Tel: +48 71 3205221
Fax: +48 71 3205273

Private address
Solskiego 19
PL - 52401 Wroclaw
Tel: +48 71 3645448

Fields of interest
Potatoes, starch, biopolymers, carbohydrates, food composition, storage

HAJDUK, Ewa Ph.D.
University of Agriculture
Department of Refrigeration and Food Concentrates
Lecturer
Podluzna 3
PL - 30239 Krakow
Tel: +48 12 4252832
Fax: +48 12 4252832

Private address
Jahody 10/1
PL - 30-348 Krakòw
Tel: +48 12 2692635

Fields of interest
Meat and meat products, poultry, proteins, freezing, storage

ICIEK, Jan Ph.D., D.Sc., Professor
Technical University of Lodz
Institute of Chemical Technology of Food
Head of Institute
Stefanowskiego 4/10
PL - 90924 Lodz
Tel: +48 42 313455
Fax: +48 42 6367488
E-Mail: i30@snack.p.lodz.pl

Private address
Tatrzanska 87M83
PL - 93279 Lodz
Tel: +48 42 6430030

Fields of interest
Sugar, quality, drying, preservation, thermal processing

IMBS, Boguslaw Professor
Food Industry Monthly Bulletin
Editor-in-Chief
Rakowiecka 36
PL - 02532 Warsaw
Tel: +48 22 8495333
E-Mail: infor.sigma@pol.pl

Private address
Przy Bazantarni 11/8
PL - 02793 Warsaw
Tel: +48 22 6491687

Fields of interest
Dairy, oils and fats, consumer research, management, quality, packaging, MAP

JARMOLUK, Andrzej Ph.D.
Agricultural University of Wroclaw
Department of Animal Products Technology
C.K. Norwida 25
PL - 50375 Wroclaw
Tel: +48 71 3205121, 3205136
Fax: +48 71 3284849
E-Mail: jar@ozi.ar.wroc.pl

Private address
Filipinska 4
PL - 52132 Wroclaw

Fields of interest
Meat and meat products, poultry, spices, additives, antioxidants, aroma active compounds, HPLC, sensory analysis, food composition, quality, preservation

KEDZIOR, Wladyslaw Dr.
University of Economics
Department of Food Commodity Science
Professor
Sienkiewicza 5
PL - 30033 Krakow
Tel: +48 12 6330821
Fax: +48 12 6335733

Private address
ul. Teligi 23/93
PL - 30835 Krakow
Tel: +48 12 4113187

Fields of interest
Meat and meat products, poultry, transgenic foods, lipids and fatty acids, sensory analysis, consumer research, food composition, nutrition, quality, quality assurance, HACCP

KMIECIK, Waldemar Andrzej Professor
Agricultural University
Department of Raw Materials and Fruit and Vegetable Processing
Professor
Podlzuna 3
PL - 30239 Krakow
Tel: +48 12 4252806
Fax: +48 12 4251801

Private address
Rysi Stok 7; PL - 30239 Krakow
Tel: +48 12 252775

Fields of interest
Fruits and vegetables, carotenoids, dietary fibres, minerals, AAS, HPLC, food composition, drying, freezing, thermal processing

KOSICKI, Zenon M.Sc., Dr.
'Celiko dr. kosicki' Slawomir Kosicki
Proxy of the owner
sw. Antoniego 71
PL - 61359 Poznan
Tel: +48 61 879376712
Fax: +48 61 8793595

Private address
sw. Antoniego 69
PL - 61359 Poznan
Tel: +48 61 8793740
Fax: +48 61 8793595

Fields of interest
Coffee, tea, cocoa, starch, carbohydrates, sensory analysis, nutrition, quality

KOSTYRA, Henryk Dr., Professor
Polish Academy of Sciences
Department of Food Chemistry
Head of Department
J. Tuwima 10
PL - 10747 Olsztyn
Tel: +48 89 5234675
Fax: +48 89 5237824
E-Mail: kos@food.pan.olsztyn.pl

Private address
Boenigka 15M4
PL - 10686 Olsztyn
Tel: +48 89 5419532

Fields of interest
Dairy, legumes, transgenic foods, Maillard reaction products, proteins, biogenic amines, electrophoresis, HPLC, sensory analysis, allergology, biotechnology

KOZLOWSKA, Halina Prof., Ph.D., D.Sc.
Institute of Animal Reproduction and Food, Research of the Polish Academy of Sciences
Director of the Institute and Head of the Department of Food Technology
Tuwima 10
PL - 10747 Olsztyn
Tel: +48 89 5234686, 5240313
Fax: +48 89 5240124
E-Mail: haka@food.irzbz.pan.olsztyn.pl

Private address
Warszawska 107
PL - 10701 Olsztyn
Tel: +48 89 5240422

Fields of interest
Fruits and vegetables, legumes, antioxidants,

carbohydrates, polyphenols, Plan Toxins, electrophoresis, GC, HPLC, sensory analysis, food composition, nutrition, biotechnology, thermal processing

KRALA, Lucjan Ph.D.
Technical University of Lodz
Department of Chemical Technology of Food
Reader
Stefanowskiego 4/10
PL - 90924 Lodz
Tel: +48 42 313463
Fax: +48 42 313402
E-Mail: lucek@snack.p.lodz.pl

Private address
Nowa 16/18M79
PL - 90031 Lodz
Tel: +48 42 749037

Fields of interest
Meat and meat products, poultry, colours, proteins, chemical reactions, food composition, quality, freezing, packaging, MAP, storage

LESNIAK, Wladyslaw Prof., Ph.D., D.Sc.
Food Biotechnology Department
Head of Department
Komandorska St. Nr. 118/120
PL - 53345 Wroclaw
Tel: +48 71 3680 260
Fax: +48 71 3672778
E-Mail: lesniak@credit.ae.wroc.pl

Private address
Lakowa St. Nr. 1
PL - 55075 Bielany Wroclawskie
Tel: +48 71 3112771

Fields of interest
Beer, beverages, spirits, additives, ethanol, sensory analysis, food composition, microbiology, biotechnology, fermentation

LESZCZYNSKI, Waclaw Ph.D., D.Sc., Professor
Agricultural University of Wroclaw
Department of Food Storage and Technology
Professor
Norwida 25
PL - 50375 Wroclaw
Tel: +48 71 3205221
Fax: +48 71 3205273

Private address
Orlowskiego 12; PL - 51637 Wroclaw
Tel: +48 71 3453136

Fields of interest
Potatoes, starch, carbohydrates, food composition, quality

LISIEWSKA, Zofia Barbara Professor
Agricultural University
Department of Raw Materials and Fruit and Vegetable Processing
Professor
Podlzuna 3
PL - 30239 Krakow
Tel: +48 12 4252806
Fax: +48 12 4251801

Private address
Rusznikarska 9/146
PL - 31261 Krakow
Tel: +48 12 6344724

Fields of interest
Fruits and vegetables, amino acids, carotenoids, dietary fibres, minerals, vitamins, AAS, HPLC, food composition, quality, drying, freezing

LISINSKA, Grazyna Ph.D., D.Sc., Professor
Agricultural University of Wroclaw
Department of Food Storage and Technology
Professor
Norwida 25
PL - 50375 Wroclaw
Tel: +48 71 3205263
Fax: +48 71 3205273

Private address
Glogowczyka 24
PL - 51604 Wroclaw
Tel: +48 71 3728520

Fields of interest
Fruits and vegetables, potatoes, carbohydrates, dietary fibres, proteins, food composition, quality, drying, storage

MATUSZEK, Tadeusz B.Sc., M.Sc. (M.Eng.), M.Sc. (Appl. Informatics), Ph.D. (in Techn. Sciences)
Technical University of Gdansk
Department of Food Engineering
Assistant Professor
G. Narutowicza Str. 11/12
PL - 80952 Gdansk
Tel: +48 58 3471674
Fax: +48 58 3415821
E-Mail: tmatusze@pg.gda.pl

Private address
Rybacka Str. 1/B7
PL - 80340 Gdansk
Tel: +48 58 5579503

Fields of interest
Food composition, fouling and cleaning, minimal

processing, packaging, MAP, rheology, thermal processing, modelling and process control, food processing, plant and machine design

MISKIEWICZ, Tadeusz Dr.habil.eng.
Academy of Economics
Food Biotechnology Department
Head of Plant Food Technology Group
Komandorska 118/120
PL - 53345 Wroclaw
Tel: +48 71 3680269
Fax: +48 71 3672778
E-Mail: tmiskiew@credit.ae.wroc.pl

Private address
Zachodina 30/13
PL - 53622 Wroclaw
Tel: +48 71 3555440

Fields of interest
Potatoes, nutrition, biotechnology, fermentation, wastewater from food industry

NOWAK, Jacek Dr.
University of Agriculture
Department of Food Technology
Assistant Professor
Wojska Polskiego 31
PL - 60624 Poznan
Tel: +48 61 8487283
Fax: +48 61 8487314
E-Mail: jacnow@ow..au.poznan.pl

Private address
Tulodziecka 29
PL - 60465 Poznan
Tel: +48 61 8422045

Fields of interest
Beer, legumes, spirits, transgenic foods, antimicrobials, ethanol, microbiology, biotechnology, fermentation, genetic engineering

PALASINSKI, Mieczyslaw M.Sc., Ing., Ph.D., D.Sc. habil., Professor
University of Agriculture
Department of Carbohydrates Technology
Professor emeritus
Al. 29-Listopada 46
PL - 31425 Krakow
Tel: +48 12 4119705
Fax: +48 12 4117753
E-Mail: rrfortuna@cyf-kr.edu.pl

Private address
ul. Baluckiego 18/1
PL - 30318 Krakow
Tel: +48 12 2661594

Fields of interest
Potatoes, starch, carbohydrates, food composition, rheology, storage

PALKA, Krystyna Dr. techn.
University of Agriculture
Department of Animal Products Technology
Lecturer on Meat Science and Technolgy
Al. 29-Listopada 52
PL - 31425 Krakow
Tel: +48 12 4119144 450
Fax: +48 12 4117753
E-Mail: rrpalka@cyf-kr.edu.pl

Private address
Podlesic 5/27
PL - 30667 Krakow

Fields of interest
Meat and meat products, poultry, proteins, preservation, rheology, thermal processing

PEKSA, Anna M.Sc., Ph.D.
Agricultural University of Wroclaw
Department of Food Storage and Technology
Lecturer
Norwida 25
PL - 50375 Wroclaw
Tel: +48 71 3205221
Fax: +48 71 3205273

Private address
Legnicka 84/8
PL - 54206 Wroclaw
Tel: +48 71 3517522

Fields of interest
Potatoes, carbohydrates, proteins, sensory analysis, food composition, rheology, storage

PEZACKI, Wincenty Full Prof., Dr., Dr. h.c. (D)
Agricultural University
Department of Meat Technology
Retired Professor
Wojska Polskiego 31
PL - 60624 Poznan
Tel: +48 61 8487263
Fax: +48 61 8487254

Private address
Mazowiezka 22
PL - 60617 Poznan
Tel: +48 61 8470517
Fax: +48 61 537053

Fields of interest
Meat and meat products, poultry, transgenic foods, water, toxic trace elements, electrophoresis, sensor analysis, sensor technology, bioavailability, consumer

research, microbiology, nutrition, quality, biotechnology, drying, fouling and cleaning, freezing, genetic engineering, high pressure technology, hygiene, packaging, MAP, quality assurance, HACCP, rheology, storage, thermal processing

PIETKIEWICZ, Jerzy Jan Ph.D., M.Sc., Eng.
Wroclow Academy of Economics
Food Biotechnology Department
Research Tutor
Komandorska 118/120
PL - 53345 Wroclaw
Tel: +48 713 680258
Fax: +48 713 672778
E-Mail: pietkiew@credit.ae.wroc.pl

Private address
Budziszynska 125/18
PL - 54436 Wroclaw
Tel: +48 713 573942

Fields of interest
Beer, beverages, sugar, honey, sugar alcohols, additives, food composition, microbiology, quality, biotechnology, fermentation, minimal processing

PRZYSIEZNA, Ewa Ph.D.
Academy of Economics
Department of Animal Food Technology
Komandorska 118/120
PL - 53345 Wroclaw
Tel: +48 71 680254
Fax: +48 71 672778

Private address
Bonczyka 10/4
PL - 51138 Wroclaw

Fields of interest
Meat and meat products, poultry, amino acids, colours, proteins, HPLC, preservation, storage

ROBAK, Malgorzata M.Sc., Ph.D.
Agricultural University of Wroclaw
Department of Biotechnology and Food Microbiology
Lecturer
Norwida 25
PL - 50375 Wroclaw
Tel: +48 71 205151
Fax: +48 71 224849
E-Mail: mrob@ozi.ar.wroc.pl

Private address
Rynek 51/14
PL - 50116 Wroclaw
Tel: +48 71 3428118

Fields of interest
Additives, aroma active compounds, biopolymers, proteins, electrophoresis, HPLC, enzymology, metabolism, microbiology, biotechnology, hygiene

RODZIEWICZ, Anna M.Sc., Ph.D.
Agricultural University of Wroclaw
Department of Biotechnology and Food Microbiology
Lecturer
Norwida 25
PL - 50375 Wroclaw
Tel: +48 71 3205116
Fax: +48 71 3224849
E-Mail: arod@ozi.ar.wroc.pl

Private address
Nabielaka 21A
PL - 51140 Wroclaw
Tel: +48 71 3255893

Fields of interest
Starch, biopolymers, carbohydrates, proteins, electrophoresis, HPLC, enzymology, microbiology, biotechnology, fouling and cleaning

RUTKOWSKI, Antoni Prof., Dr., Dr. hc. mult.
Full Member of PAS, em. ord. Professor
PaTac Kultury i Nauki p. 2102
PL - 00802 Warsaw
Tel: +48 22 260587
Fax: +48 22 6204292
E-Mail: arut@ippt.gov.pl

Private address
Marszalkowska 9/15M32
PL - 00626 Warsaw
Tel: +48 22 8251161

Fields of interest
Meat and meat products, poultry, oils and fats, additives, antioxidants, lipids and fatty acids, proteins, bioavailability, food composition, quality, preservation

RYMOWICZ, Waldemar Ph.D., D.Sc.
Agricultural University of Wroclaw
Department of Biotechnology and Food Microbiology
Lecturer
Norwida 25
PL - 50375 Wroclaw
Tel: +48 71 3205151
Fax: +48 71 3229576
E-Mail: rymowicz@ozi.ar.wroc.pl

Private address
Jesienna 6
PL - 53017 Wroclaw
Tel: +48 71 3628623
Fax: +48 71 3628623

Fields of interest
Lipids and fatty acids, trace elements, HPLC, sensor technology, biotechnology, fermentation

SCHLEGEL-ZAWADZKA, Malgorzata Ph.D.
Collegium Medicum Jagiellonian University
Department of Food Chemistry and Nutrition
Assistant Professor
Medyczna /9
PL - 30-688 Kraków
E-Mail: mfzawadz@kinga.cyf-kr.edu.pl

Private address
Bojki 5/60
PL - 30-611Kraków
Tel: +48 12 6543949
Fax: +48 12 6543949

Fields of interest
Minerals, trace elements, toxic trace elements, AAS, sensory analysis, bioavailability, consumer research, nutrition, toxicology, food preferences

SIKORA, Marek Ph.D.
University of Agriculture
Department of Carbohydrates Technology
Assistant Professor
Al. 29-Listopada 46
PL - 31425 Krakow
Tel: +48 12 4119144293
Fax: +48 12 4117753
E-Mail: rrsikora@cyf-kr.edu.pl

Private address
Kurczaba 12/37
PL - 30868 Krakow
Tel: +48 12 6572967

Fields of interest
Coffee, tea, cocoa, starch, sugar, honey, sugar alcohols, amino acids, carbohydrates, Maillard reaction products, rheology, chocolate and confectionary production

SIKORSKI, Zdzistaw Ph.D., D.Sc.
Technical University of Gdansk
Department of Food Chemistry and Technology
Head of Department
G. Narutowicza Str. 11/12
PL - 80952 Gdansk
Tel: +48 58 3471246
Fax: +48 58 34726 941
E-Mail: sikorski@chem.pg.gda.pl

Private address
Chrzanowskiego 64DM5
PL - 80278 Gdansk
Tel: +48 58 3416114

Fields of interest
Fish and fish products, meat and meat products, poultry, proteins, GC, chemical reactions, quality, freezing, preservation, storage

SKIBNIEWSKA, Krystyna A. Dr.
Warmia and Masurian University in Olsztyn
Faculty of Food Science, Institute of Commodities and Food Quality Assessment
Assistant Professor
Pl. Cieszynski 1
PL - 10957 Olsztyn
Tel: +48 89 5234966, 5234889
Fax: +48 89 5233554
E-Mail: kas@mosk.it.uwm.edu.pl

Private address
Sakowicza 4
PL - 10718 Olsztyn
Tel: +48 89 5270412

Fields of interest
Bread and cereals, minerals, trace elements, hormones, toxic trace elements, bioavailability, food composition, quality, toxicology, diet analysis

SKRABKA-BLOTNICKA, Teresa Ph.D., D.Sc.,
Professor
Academy of Economics
Department of Animal Food Technology
Head of Department
Komandorska 118/120
PL - 53345 Wroclaw
Tel: +48 71 3680254
Fax: +48 71 3672778

Private address
Sudecka 140/1
PL - 53129 Wroclaw
Tel: +48 71 3679288

Fields of interest
Meat and meat products, poultry, additives, amino acids, colours, proteins, HPLC, sensory analysis, food composition, preservation, rheology

SLOMINSKA, Lucyna Ph.D., D.Sc., Prof.
Starch and Potato Products Research Laboratory
Manager of Starch Hydrolyzates Department
Zwierzyniecka 18
PL - 60814 Poznan
Tel: +48 61 6680458
Fax: +48 61 417610
E-Mail: ls@man.poznan.pl

Private address
Osiedle Zwyciestwa 21/98
PL - 61649 Poznan
Tel: +48 61 230667

Fields of interest
Enzymatic conversion of starch, carbohydrate sweeteners

SMOLINSKA, Teresa Ph.D., D.Sc., Professor
Agricultural University of Wroclaw
Department of Animal Origin Food Technology
Norwida 25
PL - 50375 Wroclaw
Tel: +48 71 3205284
E-Mail: tsa@ozi.wroc.pl

Private address
Olszewskiego 19
PL - 51642 Wroclaw
Tel: +48 71 3487364
Fax: +48 71 3487364

Fields of interest
Eggs, meat and meat products, poultry, lipids and fatty acids, sensor technology, food composition, quality, storage

STEMPIEWICZ, Regina M.Sc., Ph.D.
Agricultural University of Wroclaw
Department of Biotechnology and Food Microbiology
Lecturer
Norwida 25
PL - 50375 Wroclaw
Tel: +48 71 3205421
Fax: +48 71 3224849
E-Mail: stemp@ozi.ar.wroc.pl

Private address
Zefirowa 1/4
PL - 53027 Wroclaw
Tel: +48 71 3398536

Fields of interest
Beer, antimicrobials, mycotoxins, TLC, metabolism, microbiology, biotechnology, fermentation

SUROWKA, Krzysztof Dr.
Agricultural University
Department of Refrigeration and Food Industry Engineering
Associate Professor
Podluzna 3
PL - 30239 Krakow
Tel: +48 12 4252832
Fax: +48 12 4252832
E-Mail: rtsurowk@cyf-kr.edu.pl

Private address
Stojalowskiego 15/16
PL - 30612 Krakow

Fields of interest
Fish and fish products, spices, antioxidants, proteins, biogenic amines, HPLC, food composition, freezing, rheology, storage

SZOLTYSEK, Katarzyna Dr.
Academy of Economics
Food Biotechnology Department
Professor
Komandorska 118/120
PL - 53345 Wroclaw
Tel: +48 71 3680271
Fax: +48 71 672778
E-Mail: catherin@credit-ae.wroc.pl

Private address
Wapienna 15/1
PL - 50518 Wroclaw
Tel: +48 71 3360583

Fields of interest
Bread and cereals, cosmetics, oils and fats, sweeteners, additives, carotenoids, lipids and fatty acids, biogenic amines, electrophoresis, HPLC, sensory analysis, sensor technology, consumer research, enzymology, quality, biotechnology, genetic engineering, quality assurance, HACCP

TOMASIK, Piotr Professor, Ph.D., D.Sc.
University of Agriculture
Department of Chemistry
Professor
Mickiewicz Ave. 21
PL - 31120 Krakow
Tel: +48 12 338826
Fax: +48 12 336245
E-Mail: rrtomasi@cyf-kr.edu.pl

Private address
Os. 2 Pulku Lotniczego 9/151
PL - 31867 Krakow
Tel: +48 12 6473660

Fields of interest
Starch, biopolymers, carbohydrates, trace elements, spectroscopy, thermal anlaysis, chemical reactions, freezing, high pressure technology, rheology

TROJAN, Marek Ph.D., D.Sc.
Annuitant

Private address
Pogorzelska 6/8
PL - 60162 Poznan
Tel: +48 61 8685291

Fields of interest
Eggs, meat and meat products, poultry, lipids and fatty acids, minerals, AAS, chemometrics, electrophoresis, sensory analysis, sensor technology

TROJANOWSKA, Krystyna M.Sc., Ph.D.
Agricultural University
Department of Biotechnology and Food Microbiology
Research Worker
Mazowiecka 48
PL - 60623 Poznan
Tel: +48 61 8487355

Private address
Dzialowa 10M26
PL - 61747 Poznan
Tel: +48 61 8527293

Fields of interest
Antimicrobials, vitamins, microbiology,
biotechnology, hygiene

WILSKA-JESZKA, Jadwiga D.Sc., Professor
Technical University of Lodz
Department of Technical Biochemistry
Professor
Stefanowskiego 4/10
PL - 90924 Lodz
Tel: +48 42 6313435
Fax: +48 42 6313402
E-Mail: jkjeszka@pdi.lodz.pl

Private address
Pomorska 130M38
PL - 91403 Lodz
Tel: +48 42 6789796

Fields of interest
Beverages, fruits and vegetables, legumes,
Antioxidants, colours, oxidation reactions,
polyphenols, vitamins, HPLC, food composition

WITKOWSKA, Danuta Ph.D., D.Sc.
Agricultural University of Wroclaw
Department of Biotechnology and Food Microbiology
Reader
Norwida 25
PL - 50375 Wroclaw
Tel: +48 71 3205114
Fax: +48 71 3224849
E-Mail: wit@ozi.ar.wroc.pl

Private address
Damrota 35/15
PL - 50306 Wroclaw
Tel: +48 71 3213131

Fields of interest
Aroma active compounds, biopolymers,
carbohydrates, proteins, electrophoresis, HPLC,
enzymology, metabolism, microbiology,
biotechnology

WOJTATOWICZ, Maria Ph.D., D.Sc.
Agricultural University of Wroclaw
Department of Biotechnology and Food Microbiology
Reader
Norwida 25
PL - 50375 Wroclaw
Tel: +48 71 3205117
Fax: +48 71 3224849
E-Mail: mwojt@ozi.ar.wroc.pl
Private address
Bacciarellego 40B/14
PL - 51649 Wroclaw
Tel: +48 71 3451291

Fields of interest
Dairy, additives, Antimircobials, microbiology,
biotechnology, fermentation, preservation

ZEGOTA, Henryk Dr.
Technical University of Lodz
Department of Applied Radiation Chemistry
Head of Radiation Food Chemistry Group
Wroblewskiego 15
PL - 93590 Lodz
Tel: +48 42 6313183
Fax: +48 42 6360246

Private address
J. Przybosia 48
PL - 91170 Lodz
Tel: +48 42 6560898

Fields of interest
Fruits and vegetables, spices, antioxidants,
biopolymers, carbohydrates, Maillard reaction
products, oxidation reactions, vitamins, mycotoxins,
HPLC, analytical chemistry, food composition,
irradiation, preservation

ZIAJKA, Stefan Ph.D., M.Sc.
University of Agriculture and Technology
Institute of Dairy Science and Technology
Development
Professor
Oczapowskiego 7
PL - 10957 Olsztyn
Tel: +48 89 233377
Fax: +48 89 273908
E-Mail: zjajka@moskit.art.olsztyn.pl

Private address
Sloneczna 20
PL - 107010 Olsztyn
Tel: +48 89 5240289

Fields of interest
Dairy, allergology, quality, hygiene, quality assurance
HACCP

ZYLA, Krzysztof Ph.D.
University of Agriculture
Department of Food Biotechnology
Associate Professor, Department Head
Al. 29-Listopada 46
PL - 31425 Krakow
Tel: +48 12 4119144576
Fax: +48 12 4117753
E-Mail: rrzyla@cyf-kr.edu.pl

Private address
Rozrywka 20/5
PL - 31419 Krakow
Tel: +48 12 9387

Fields of interest
Bread and cereals, legumes, meat and meat products, poultry, transgenic foods, antioxidants, carbohydrates, dietary fibres, minerals, proteins, AAS, electrophoresis, HPLC, spectroscopy, bioavailability, enzymology, microbiology, nutrition, biotechnology, fermentation

Portugal

AMARO PINTO, Rui Manuel Graduate in Pharmacy, Quality Control and Food Technology Master
Faculty of Pharmacy of Lisbon
Assistant of Pharmacology
Av. Forcas Armadas
P - 1600 Lisboa
Tel: +351 21 7946400, 7946422
Fax: +351 21 7957458

Private address
Av. Marista N. 5, 8A
P - 2700 Parede

Fields of interest
Bread and cereals, water, minerals, steroids, trace elements, electrochemistry, electrophoresis, PCR, spectroscopy, toxicology

COIMBRA, Manuel António B.Sc., Ph.D.
University of Aveiro
Department of Chemistry
Professor
Campus Universitario de Santiago
P - 380-193 Aveiro
Tel: +351 234 37036023521
Fax: +351 234 370760
E-Mail: mac@dq.ua.pt

Private address
Rua Ciudad Rodrigo 5, 3Esq
P - 3810 Aveiro
Tel: +351 234 426619

Fax: +351 234 370706

Fields of interest
Coffee, tea, cocoa, fruits and vegetables, legumes, wine, aroma active compounds, biopolymers, carbohydrates, dietary fibres, GC, food composition

DELGADILLO, Ivonne Dr.rer.nat, Prof.
University of Aveiro
Department of Chemistry
Lecturer, Head of Food Biochemistry Group
Campo Universitário
P - 3810 Aveiro
Tel: +351 234 370718
Fax: +351 234 370084
E-Mail: ivonne@dq.ua.pt

Private address
Rua das Cerejeiras 17
P - 3800 Aveiro
Tel: +351 234 316020

Fields of interest
Fruits and vegetables, wine, aroma active compounds, carbohydrates, proteins, sensor technology, spectroscopy, adulteration, quality

DOS SANTOS BAPTISTA, Bráulio M.Sc.
ISEL (Lisbon Engineering Superior Institute)
Teacher
Rua Conselheiro Emítio Navarro
P - 1900 Lisboa
Tel: +351 1 8590215109
Fax: +351 1 8597046

Private address
Praeeta Anelar Brotere 11 - 3 Fr
P - 2830 Barreiro
Tel: +351 1 2152193

Fields of interest
Cosmetics, wine, antioxidants, aroma active compounds, plant toxins, GC, chemical reactions, toxicology, drying, high pressure technology

EMPIS, José Ph.D., Prof.
Instituto Superior Técnico - Dept. Eng. Quimica
Professor
Avenida Rovislo Pais
P - 1049-001 Lisboa
Tel: +351 21 8417889
E-Mail: pcjempis@alfa.ist.utl.pt, jose-empis@hotmail.com

Private address
Tel: +351 91 9527247

Fields of interest
Fish and fish products, oils and fats, antioxidants,

Portugal

carotenoids, oxidation reactions, alkaloids, spectroscopy, biotechnology, irradiation, minimal processing

---

GIL, Ana Maria Pissarra C. Ph.D.
University of Aveiro, Department of Chemistry
Associate Professor
Campus de Santiago
P - 3800 Aveiro
Tel: +351 34 37036023522
Fax: +351 34 370084
E-Mail: agil@dq.ua.pt

Private address
Av. Forca Aerea, 40
P - 3800 Aveiro
Tel: +351 34 383292

Fields of interest
Beverages, bread and cereals, fish and fish products, fruits and vegetables, biopolymers, carbohydrates, proteins, chemometrics, sensory analysis, spectroscopy

---

MOREIRA DA SILVA, Aida Maria M.Sc., Ph.D.
Escola Superior Agrária
Professora Adjunta
Bencanta
P - 3000 Coimbra
Tel: +351 39 802940
Fax: +351 39 802979

Private address
Rua Dos Combatentes 130/5
P - 3030 Coimbra
Tel: +351 39 717619
Fax: +351 39 26541

Fields of interest
Water, carbohydrates, spectroscopy, enzymology, food composition, biotechnology

---

ROGERSON, Frank B.Sc., M.Sc., Ph.D.
Universidade do Porto
Departmento de Quimica, Faculdade de Ciências
Post-Doctorate Researcher
Rua do Campo Alegre 687
P - 4169-700 Porto
Tel: +351 2 26082867
E-Mail: steverogerson@mail.tekpac.pt

Private address
Rua Antoni Ferreira da costa Maia 61-1
P - 4470-137 Porto
Tel: +351 2 29410020

Fields of interest
Spirits, wine, aroma active compounds, carotenoids, sensory analysis, chemical reactions, enzymology, food composition, fermentation, GC, MS

SERENO, Alberto M.C. Chem.Eng., M.Sc., Ph.D.
University of Porto
Faculty of Engineering
Associate Professor
Rua Bragas
P - 4040-123 Porto
Tel: +351 22 2041655
Fax: +351 22 2000808
E-Mail: sereno@fe.up.pt

Private address
Tv. Simoes Almeida 29
P - 4460 Custoias Mts
Tel: +351 22 9534743

Fields of interest
Fruits and vegetables, legumes, spices, spectroscopy, thermal anlaysis, drying, freezing, minimal processing

---

SOUSA, Isabel Prof., Dr.
Universidade Tecnica de Lisboa
Instituto Superior de Agronomia
Lecturer
Tapada da Ajuda
P - 1349-017 Lisboa
Tel: +351 1 3602040
Fax: +351 1 3635031
E-Mail: isabelsousa@cipop.isl.utl.pt

Fields of interest
Carbohydrates, proteins, sensory analysis, food composition, rheology

Slovak Republic

FARKAS, Pavel Ph.D.
Food Research Institute
Research Scientist
Priemyselná 4, P.O.Box 25
SK - 82475 Bratislava
Tel: +421 7 50237112
Fax: +421 7 55571417
E-Mail: pavel.farkas@vup.sk

Private address
Zehrianska 7
SK - 85107 Bratislava
Tel: +421 7 63821122

Fields of interest
Meat and meat products, poultry, spices, Spiritis, wine, additives, aroma active compounds, GC, MS,

adulteration, biotechnology

KOVAC, Milan Dipl.-Ing., Ph.D.
Food Research Institute
Director of Institute
Priemyselná 4, P.O.Box 25
SK - 82475 Bratislava
Tel: +421 7 55574622
Fax: +421 7 55571417
E-Mail: milan.kovac@vup.sk

Private address
Viglasska 4
SK - 85106 Bratislava
Tel: +421 7 63828354

Fields of interest
Additives, antioxidants, aroma active compounds, colours, adulteration, analytical chemistry, chemical reactions, food composition, management, quality assurance, HACCP

KRKOSKOVA, Bernadetta Ing., CSc.
Food Research Institute
Head of Technology Department
Priemyselná 4, P.O.Box 25
SK - 82475 Bratislava
Tel: +421 7 50237168
Fax: +421 7 55571417
E-Mail: krkoskova@vup.sk

Private address
Stanekova 4
SK - 84103 Bratislava
Tel: +421 7 64789237
Fax: +421 7 64789237

Fields of interest
Bread and cereals, legumes, meat and meat products, poultry, additives, carbohydrates, dietary fibres, fat replacers, electrophoresis, TLC, quality

MRAZ, Igor Dipl.-Ing., Ph.D.
LikoSpol a.s.
Director
Mileticova 23, P.O. Box 4
SK - 82462 Bratislava 26
Tel: +42 7 55574831, 55577191, 50244419
Fax: +42 7 55425480
E-Mail: likospol@isternet.sk

Private address
Karadzicova 55
SK - 81107 Bratislava 1
Tel: +42 7 55561253

Fields of interest
Fruits and vegetables, meat and meat products, poultry, potatoes, additives, quality, minimal processing, packaging, MAP, preservation, quality assurance, HACCP, themal processing

SCHMIDT, Stefan Dr., Ph.D., Assoc. Prof.
Slovak Technical University
Department of Milk, Fat and Food Hygiene
Head of Lipid Technology Section
Radlinskeho 9
SK - 81237 Bratislava
Tel: +42 17 59325434
Fax: +42 17 52493198
E-Mail: schmidt@checdek.chtf.stuba.sk

Private address
Mlynarovicova 24
SK - 85104 Bratislava
Tel: +42 7 812160

Fields of interest
Oils and fats, antioxidants, lipids and fatty acids, polyphenols, GC, HPLC, thermal anlaysis, TLC, chemical reactions, quality, storage

SIMKO, Peter Assoc. Prof., Ph.D.
Food Research Institute
Vice Director of Food Research
Priemyselná 4, P.O.Box 25
SK - 82475 Bratislava
Tel: +421 7 55574622
Fax: +421 7 55571417
E-Mail: peter.simko@vup.sk

Private address
Kadnárova 2/17
SK - 83105 Bratislava
Tel: +421 7 44881888

Fields of interest
Meat and meat products, poultry, additives, colours, PAHs, GC, HPLC, packaging, MAP

SINKOVÁ, Terèzia Ing., C.Sc.
Food Research Institute
Head of Department
Priemyselná 4, P.O.Box 25
SK - 82475 Bratislava
Tel: +421 7 50237150
Fax: +421 7 5551417

Private address
K. Adlera 36
SK - 84102 Bratislava
Tel: +421 7 6482986
Fax: +421 7 64282986

Fields of interest
Additives, quality, legislation

SUHAJ, Milan Ing., CSc.
Food Research Institute
Head of analytical Department
Priemyselná 4, P.O.Box 25
SK - 82475 Bratislava
Tel: +421 7 50237146
Fax: +421 7 55571417
E-Mail: milan.suhaj@vup.sk

Private address
Prokopa Velkeho 41
SK - 81707 Bratislava 1

Fields of interest
Beverages, oils and fats, sweeteners, additives, antioxidants, fat replacers, plant toxins, CE, spectroscopy, adulteration, analytical chemistry, food composition, nutrition, quality, toxicology, quality assurance

Slovenia

CERKVENIK, Vesna M.Sc.
University of Ljubljana
Veterinary Faculty, Institute for Food Hygiene and Bromathology
Research Chemist
Gerbiceva 60
SLO - 1000 Ljubljana
Tel: +386 61 1779100
Fax: +386 61 332243

Private address
Chengdujska 22
SLO - 1000 Ljubljana
Tel: +386 61 1409668

Fields of interest
Dairy, eggs, fish and fish products, meat and meat products, poultry, antimicrobials, steroids, drugs, hormones, pesticides, GC, HPLC, analytical chemistry, consumer research

DOGANOC, Darinka Zdenka Ph.D.
University of Ljubljana
Veterinary Faculty, Department of Food Hygiene
Ass. Professor
Gerbiceva 60
SLO - 1000 Ljubljana
Tel: +386 61 1779188
Fax: +386 61 322243

Private address
Rozna Dolina C.III/8
SLO - 1000 Ljubljana
Tel: +386 61 742549

Fields of interest
Dairy, eggs, fish and fish products, meat and meat products, poultry, trace elements, toxic trace elements, AAS, food composition, quality, hygiene

EKAR, Igor Univ. Dipl.-Ing. Chem.
Institute of Public Health Kranj
Head of Laboratory
Gosposvetska 12
SLO - 64000 Kranj
Tel: +386 64 282659
Fax: +386 64 226702
E-Mail: igor.ekar@gov.si

Private address
Cesta 1. Maja 73
SLO - 64000 Kranj
Tel: +386 64 324567

Fields of interest
Cosmetics, dairy, oils and fats, spices, sugar, honey, sugar alcohols, additives, antioxidants, colours, lipids and fatty acids, HPLC

GLADOVIC, Natasa B.Sc. chem.
Droga d.o.o. Portoroz
Senior Scientist
Seca 112
SLO - 66320 Portoroz
Tel: +386 66 771785
Fax: +386 66 771782

Private address
Vrtna 1
SLO - 66310 Izola
Tel: +386 66 275748

Fields of interest
HPLC, food composition, nutrition, quality, HACCP

GOLJA, Viviana Dipl.-Ing. chem., M.Sc.
Institute of Public Health of the Republic of Slovenia
Coordinator
Grabloviceva 44
SLO - 61000 Ljubljana
Tel: +386 61 1402030 396
Fax: +386 61 1403379
E-Mail: viviana.golja@gov.si

Private address
Moskriceva 36
SLO - 61000 Ljubljana
Tel: +386 61 444828

Fields of interest
Cosmetics, additives, spectroscopy, analytical chemistry, regulative issues, packaging, MAP

RIHAR JEREB, Bernarda M.Sc.
Kolinska
Technologist in R and D Department
Smartinska C. 30
SLO - 61000 Ljubljana
Tel: +386 61 1721 500
Fax: +386 61 1721571

Private address
Polhov Gradec 1/c
SLO - 61355 Polhov Gradec
Tel: +386 61 645096

Fields of interest
Fruits and vegetables, fermentation, filtration

SINIGOJ-GACNIK, Ksenija Dr.
University of Ljubljana
Veterinary Faculty
Research Associate
Gerbiceva 60
SLO - 1000 Ljubljana
Tel: +386 61 1779187
Fax: +386 61 332243
E-Mail: gacnikks@mail.vf.uni-lj.si

Private address
Krivec 56
SLO - 1000 Ljubljana
Tel: +386 61 1599648

Fields of interest
Dairy, eggs, fish and fish products, meat and meat products, poultry, additives, antimicrobials, trace elements, drugs, toxic trace elements, AAS, HPLC

UKMAR MALJEVAC, Damjana B.Sc. chem.
Droga d.o.o. Portoroz
Head of Laboratory
Seca 112
SLO - 6320 Portoroz
Tel: +386 66 771360
Fax: +386 66 771782

Private address
II. Prekomorske Brigade 10A
SLO - 6000 Koper
Tel: +386 66 275160

Fields of interest
HPLC, food composition, nutrition, quality, quality assurance, HACCP

ZIDARIC, Metka B.Sc.
Public Health Institute, Environmental Protection Institute
Prvomajska 1
SLO - 2000 Maribor
Tel: +386 62 4500100

Fax: +386 62 413978

Private address
Naversnikova 32
SLO - 2000 Maribor
Tel: +386 62 631734

Fields of interest
Beer, coffee, tea, cocoa, fruits and vegetables, meat and meat products, poultry, oils and fats, additives, PCBs, pesticides, toxic trace elements, food composition

ZORIC, Andreja Dipl.-Ing. chem.
Institute of Public Health of the Republic of Slovenia
Analyst
Grabloviceva 44
SLO - 1000 Ljubljana
Tel: +386 61 1402030396
Fax: +386 61 1403379

Private address
Kebetova 1
SLO - 61215 Medvode

Fields of interest
Cosmetics, additives, antioxidants, HPLC, analytical chemistry, regulative issues, packaging, MAP

Spain

AUBOURG, Santiago Dr.
Instituto de Investigaciones Marinas (CSIC)
Investigador Cientifico
Eduardo Cabello 6
E - 36208 Vigo
Tel: +34 86 231930 233
Fax: +34 86 292762
E-Mail: saubourg@nautilus.iim.csic.es

Private address
Manuel Cominges, 64 48
E - 36213 Vigo
Tel: +34 86 293544
Fax: +34 86 293544

Fields of interest
Fish and fish products, antioxidants, lipids and fatty acids, oxidation reactions, spectroscopy, food composition, quality, freezing, thermal processing

AUBOURG, Santiago Dr.
Instituto de Investigaciones Marinas (CSIC)
Investigador Cientifico
Eduardo Cabello 6

E - 36208 Vigo
Tel: +34 86 231930 233
Fax: +34 86 292762
E-Mail: saubourg@nautilus.iim.csic.es

Private address
Manuel Cominges, 64
E - 36213 Vigo
Tel: +34 86 293544
Fax: +34 86 293544

Fields of interest
Fish and fish products, antioxidants, lipids and fatty acids, oxidation reactions, spectroscopy, food composition, quality, freezing, thermal processing

---

BARBER, Salvador Dr., Food Science and Technology Diplomate
Espanola de I + D, S.A.
Director
Poligono Virgen de los Dolores C 215-209, s/n
E - 46113 Moncada (Valencia)
Tel: +34 96 1394461
Fax: +34 96 1394026
E-Mail: espanid@teleline.es

Private address
Urbanización Cruz de Garcia, Palmeras 14
E - 46100 Burjasot (Valencia)
Tel: +34 96 3637939

Fields of interest
Research and development

BARBER, Berta Dr., Food Science and Technology Diplomate
Espanola de I + D, S.A.
Development of Processes
Poligono Virgen de los Dolores C 215-209, s/n
E - 46113 Moncada (Valencia)
Tel: +34 96 1394461
Fax: +34 96 1394026
E-Mail: espanid@teleline.es

Private address
Guillem de Castro 51-12
E - 46007 Valencia
Tel: +34 96 3511685

Fields of interest
Bread and cereals, additives, research and development

BARRON, Luis Javier Rodríguez Dr.
Universidad del Pais Vasco
Departamento Technologia de Alimentos
Professor (as a Lecturer)
Paseo de la Universidad 7
E - 01006 Vitoria-Gasteiz

Tel: +34 945 013082
Fax: +34 945 130756
E-Mail: knprobal@vf.ehu.es

Fields of interest
Dairy, oils and fats, aroma active compounds, lipids and fatty acids, GC, HPLC, TLC, food composition, biotechnology, fermentation

---

BENEDITO, Carmen Ph.D.
Instituto de Agroquimica y Tecnologia de Alimentos, CSIC
Professor of Research
P.O. Box 73
E - 46100 Burjasot Valencia
Tel: +34 96 3900022
Fax: +34 96 3636301
E-Mail: cbenedito@iata.csic.es

Private address
Palmeras 14
E - 46980 Paterna
Tel: +34 96 3637939

Fields of interest
Bread and cereals, carbohydrates, HPLC, spectroscopy, thermal anlaysis, chemical reactions, food composition, quality, fermentation, rheology

---

CALVO, Marta Maria BVM, Ph.D.
Instituto de Fermentaciones Industriales, CSIC
Tenured Scientist
Juan de la Cierva 3
E - 28006 Madrid
Tel: +34 1 5622900240
Fax: +34 1 5644853
E-Mail: mmcalvo@ifi.csic.es

Private address
C/Jarama 6
E - 28220 Majadahonda (Madrid)
Tel: +34 91 5885200

Fields of interest
Beverages, dairy, carbohydrates, proteins, electrophoresis, GC, HPLC, sensory analysis, TLC, adulteration, allergology, analytical chemistry, food composition, microbiology, nutrition, fermentation, filtration, storage, thermal processing

---

CATALÁ, Ramon Ph.D.
Instituto de Agronomica y Tecnologia de Alimentos (CSIC)
Research Professor, Head of Department of Food Quality and Preservation
Apartado de Correos 73
E - 46100 Burjasot Valencia
Tel: +34 6 3900022

Fax: +34 36 3636301
E-Mail: rcatala@iata.csic.es

Private address
Cervantes 31
E - 46183 L' Eliana (Valencia)
Tel: +34 96 2743970

Fields of interest
Fruits and vegetables, packaging, MAP

COLLAR ESTEVE, Concha Ph.D.
Instituto de Agroquimica y Tecnologia de Alimentos (CSIC)
Researcher
Poligono la Coma s/n
E - 46980 Paterna (Valencia)
Tel: +34 96 3900022 2106
Fax: +34 96 3636301

Private address
Avda. Pérez Galdós 64
E - 46008 Valencia
Tel: +34 96 3824780, 3637757
Fax: +34 96 3824780

Fields of interest
Bread and cereals, additives, amino acids, lipids and fatty acids, proteins, electrophoresis, GC, HPLC, thermal anlaysis, chemical reactions, fermentation, rheology, storage, thermal processing

DURAN, Luis M.Sc., Dr.
Instituto de Agroquímica y Tecnología de Alimentos (CSIC)
Research Professor
Poligono de la Coma s/n
E - 46980 Paterna (Valencia)
Tel: +34 963 900022
Fax: +34 963 636301

Private address
Avenida de Aragon 36/1-19
E - 46021 Valencia
Tel: +34 963 694240

Fields of interest
Fruits and vegetables, sweeteners, carbohydrates, sensory analysis, quality, rheology, texture measurement

FRIAS, Juana B.Sc., Ph.D.
Instituto de Fermentaciones Industriales
Tenured Scientist
Juan de la Cierva 3
E - 28006 Madrid
Tel: +34 1 5622900357
Fax: +34 1 5644853

Fields of interest
Dairy, fruits and vegetables, legumes, antioxidants, carbohydrates, dietary fibres, vitamins, plant toxins, CE, chemometrics, HPLC, analytical chemistry, bioavailability, food composition, biotechnology, fermentation, thermal processing, germination

GONZALEZ-SANJOSE, Maria Luisa Ph.D.
Burgos University
Faculty of Science
Professor
Plaza Misael Banuelos S/N
E - 09001 Burgos
Tel: +34 47 258815
Fax: +34 47 258852
E-Mail: marglez@ubu.es

Private address
Avda. CID, N. 2
E - 09005 Burgos

Fields of interest
Fruits and vegetables, wine, antioxidants, colours, polyphenols, sensory analysis, consumer research, food composition, quality, fermentation

GUILLEN, M. D. Dr.
Universidad del Pais Vasco, Facultad de Farmacia
Tecnologia de los Alimentos
Paseo de la Universidad N. 7
E - 1006 Vitoria
Tel: +34 945 013081
Fax: +34 945 130756
E-Mail: knpgulod@vf.ehu.es

Fields of interest
Dairy, fish and fish products, meat and meat products, poultry, oils and fats, spices, antimicrobials, antioxidants, aroma active compounds, lipids and fatty acids, PAHs, chemometrics, GC, MS, spectroscopy, adulteration, analytical chemistry, food composition, quality, preservation

HEREDIA, Antonia Dr.
Instituto de la Grassa
Avda. Padre Garcia Tejero 4
E - 41012 Sevilla
Tel: +34 54 691054
Fax: +34 54 691262
E-Mail: ahmoreno@cica.es

Private address
San Vicente de Paúl 10
E - 4101 Sevilla
Tel: +34 54 343982

Fields of interest
Fruits and vegetables, oils and fats, carbohydrates, dietary fibres, polyphenols, electrophoresis, GC,

HPLC, food composition, biotechnology

HERRAIZ TOMICO, Tomas Ph.D.
Consejo Superior de Investigaciones Cientificas
Instituto de Fermentaciones Industriales
Research
Juan de la Cierva 3
E - 28006 Madrid
Tel: +34 91 5622900361
Fax: +34 91 5644853
E-Mail: ijiht16@ifi.csic.es

Private address
Pitagoras 4
E - 28224 Las Rozas (Madrid)
Tel: +34 91 6317281

Fields of interest
Amino acids, alkaloids, heterocyclic aromatic amines, GC, HPLC, MS, analytical chemistry, chemical reactions, food composition, toxicology

JUAREZ, Manuela Dr.
Instituto del Frio (CSIC)
Research Professor
Ciudad Universitaria
E - 28040 Madrid
Tel: +34 91 5492300
Fax: +34 91 5493627
E-Mail: mjuarez@if.csic.es

Private address
Azalea 285
E - 28109 Alcobendas (Madrid)
Tel: +34 91 6502338

Fields of interest
Dairy, oils and fats, lipids and fatty acids, minerals, trace elements, GC, adulteration, analytical chemistry, food composition, quality

LLACER, Dolores Dr., Food Science and Technology Diplomate
Espanola de I + D, S.A.
Food Products Development
Poligono Virgen de los Dolores C 215-209, s/n
E - 46113 Moncada (Valencia)
Tel: +34 96 1394461
Fax: +34 96 1394026
E-Mail: espanid@teleline.es

Private address
Pintor Ferrer Calatayud 29; E - 46022 Valencia
Tel: +34 96 3717674

Fields of interest
Bread and cereals, additives, research and development

MARTINEZ ANAYA, Antonia M. Ph.D.
Instituto de Agroquimica y Tecnologia de Alimentos (CSIC)
Researcher
P.O. Box 73
E - 46100 Burjasot Valencia
Tel: +34 6 3900022
Fax: +34 6 3636301

Private address
Torres-Torres 18
E - 46018 Valencia
Tel: +34 6 3841188

Fields of interest
Bread and cereals, additives, carbohydrates, chemometrics, HPLC, thermal anlaysis, quality, fermentation, rheology

MARTINEZ-CASTRO, Isabel Ph.D.
Consejo Superior de Investigaciones Cientificas
Instituto de Quimica Organica
Research
Juan de la Cierva 3
E - 28006 Madrid
Tel: +34 1 5622900 212
Fax: +34 1 5644853
E-Mail: igomc16@iqog.csic.es

Private address
Jorge Juan 36
E - 28001 Madrid

Fields of interest
Dairy, sugar, honey, sugar alcohols, aroma active compounds, carbohydrates, Maillard reaction products, GC, MS, food composition

OLANO, Agustin B.Sc., Ph.D.
Instituto de Fermentaciones Industriales, CSIC
Vice-Director
Juan de la Cierva 3
E - 28006 Madrid
Tel: +34 1 5622900242
Fax: +34 1 5644853

Fields of interest
Dairy, fruits and vegetables, amino acids, carbohydrates, Maillard reaction products, GC, HPLC, analytical chemistry, chemical reactions, food composition, high pressure technology

ORTOLA, Concepcion Dr., Food Science and Technology Diplomate
Espanola de I + D, S.A.
Food Products Development
Poligono Virgen de los Dolores C 215-209, s/n
E - 46113 Moncada (Valencia)
Tel: +34 96 1394461

Fax: +34 96 1394026
E-Mail: espanid@teleline.es

Private address
Blasco Ibanez, no 119-35A
E - 46022 Valencia
Tel: +34 96 3727015

Fields of interest
Bread and cereals, additives, research and development

PASTORIZA, Laura Dr.
Instituto de Investigaciones Marinas de Vigo del Consejo Superior de Investigeciones Cientificas (Marine Research Institute)
Head of Research Group
Eduardo Cabello 6
E - 36208 Vigo
Tel: +34 86 231930209
Fax: +34 86 292762
E-Mail: laura@iim.csic.es

Private address
Via Hispanidad 90/1
E - 36203 Vigo
Tel: +34 86 472214

Fields of interest
Fish and fish products, additives, antioxidants, carbohydrates, thermal anlaysis, food composition, microbiology, packaging, MAP, preservation, storage

---

REGLERO, Guillermo J. Ph.D.
Universidad Autonoma de Madrid
Tecnologia de Alimentos
Professor
Carretera de Colmenar 15
E - 28049 Madrid
Tel: +34 91 3978128
Fax: +34 91 3978255
E-Mail: guillermo.reglero@uam.es

Private address
Avda Independencia 39
E - 28700 S.S. Reyes (Madrid)

Fields of interest
Fruits and vegetables, spices, antioxidants, aroma active compounds, extraction (supercritical fluids)

VIDAL-VALVERDE, Concepcion B.Sc., Ph.D.
Instituto de Fermentaciones Industriales, CSIC
Research Professor
Juan de la Cierva 3
E - 28006 Madrid
Tel: +34 91 5622900241
Fax: +34 91 5644853
E-Mail: ificv12@ifi.csic.es

Private address
Torrelaguna 123
E - 28043 Madrid
Tel: +34 91 4166286

Fields of interest
Legumes, antioxidants, carbohydrates, dietary fibres, vitamins, electrochemistry, HPLC, bioavailability, food composition, nutrition, biotechnology, fermentation

Sweden

JÄGERSTAD, Margaretha Ph.D., Prof.
Department of Food Science, Division of Food Chemistry
Head of Food Chemistry
Undervisningsplan 6C, Box 7051
S - 75007 Uppsala
Tel: +46 18 671991
Fax: +46 18 672995
E-Mail: magaretha.jagerstad@lmv.slu.se

Private address
Entitvägen 7
S - 75652 Uppsala
Tel: +46 18 321863

Fields of interest
Bread and cereals, dairy, fruits and vegetables, meat and meat products, poultry, Maillard reaction products, vitamins, heterocyclic aromatic amines, GC, HPLC, thermal anlaysis, thermal processing

---

LINGNERT, Hans Ph.D.
The Swedish Institute for Food and Biotechnology
Research Director
P.O. Box 5401
S - 40229 Göteborg
Tel: +46 31 355652
Fax: +46 31 833782
E-Mail: nl@sik.de

Private address
Hakansdal 24
S - 41749 Göteborg

Fields of interest
Antioxidants, lipids and fatty acids, Maillard reaction products, oxidation reactions, sensory analysis, chemical reactions, management, quality, storage

---

ÖSTERDAHL, Bengt-Göran Ph.D.
National Food Administration
Head of Chemistry Division 1
P.O. Box 622
S - 75126 Uppsala

Tel: +46 18 175500
Fax: +46 18 105848
E-Mail: bgos@slv.se, b-g@delta.telenordia.se

Private address
Lagundavägen 7
S - 74082 Örsundsbro
Tel: +46 171 460141

Fields of interest
Fruits and vegetables, meat and meat products, poultry, water, drugs, pesticides, GC, HPLC, MS, analytical chemistry, management

---

SANDBERG, Ann-Sofie Ph.D, Professor
Chalmers University of Technology
Department of Food Science
Head of Department
P.O. Box 5401
S - 40229 Göteborg
Tel: +46 31 355630
Fax: +46 31 833782
E-Mail: ann-sofie.sandberg@chalmers.se

Private address
Alfhöjdsgatan 16
S - 43138 Mölndal

Fields of interest
Bread and cereals, legumes, dietary fibres, minerals, trace elements, HPLC, bioavailability, nutrition, fermentation

Switzerland

BATTAGLIA, Reto Dr.Sc.techn.ETH, dipl.chem.ETH, eidg. dipl. Lebensmittelchemiker
Migros-Genossenschafts-Bund
Director, Head of Migros Scientific Services
CH - 8031 Zürich
Tel: +41 1 2773140
Fax: +41 1 2773170
E-Mail: reto.battaglia@mgb.ch

Private address
Talstr. 28
CH - 8634 Hombrechtikon
Tel: +41 55 2442809

Fields of interest
Adulteration, analytical chemistry, bioavailability, food composition, management, microbiology, nutrition, quality, regulative issues, toxicology

---

BOSSET, Jacques Olivier Ph.D.
Swiss Federal Dairy Research Station
Senior Analyst / Head of the group 'Metabolites'
Schwarzenburgstr. 161
CH - 3003 Bern
Tel: +41 31 3238167
Fax: +41 31 3238227
E-Mail: jacques-olivier.bosset@fam.admin.ch

Fields of interest
Dairy, aroma active compounds, GC, MS, sensor technology, analytical chemistry, food composition, fermentation, packaging, MAP, storage

---

CORVI, Claude Albert Docteur ès Sciences - mention chimie
Service du chimiste cantonal
Head Manager
22, Quai Ernest-Ansermet
CH - 1205 Geneva
Tel: +41 22 3287511
Fax: +41 22 3280150
E-Mail: claude.corvi@etat.ge.ch

Private address
243, Chemin des Volandes
F - 74380 Cranves - Sales
Tel: +33 50 393080

Fields of interest
PCBs, pesticides, toxic trace elements, AAS, GC, HPLC, food composition, toxicology, hygiene

---

GALLUSER, Anita Dr.rer.nat.
Bischofszell Nahrungsmittel AG
Head of Quality Assurance
Nordstr.
CH - 9220 Bischofszell
Tel: +41 71 4249111
Fax: +41 71 4249499
E-Mail: anita.galluser@bina.ch

Private address
Sandackerstr. 5; CH - 9245 Oberbüren
Tel: +41 71 9513744

Fields of interest
Beverages, coffee, tea, cocoa, fruits and vegetables, legumes, meat and meat products, poultry, potatoes, additives, vitamins, management, quality, hygiene, quality assurance, HACCP

GAUDARD - DE WECK, Daniele Ph.D.
Nestlé Research Centre
Senior Scientist
P.O. Box 44
CH - 1000 Lausanne 26
Tel: +41 21 7858793
Fax: +41 21 7858556

Private address
Place du Village /La Carree

CH - 1041 Bretigny - sur - Morrens
Tel: +41 21 7314523

Fields of interest
Nutrition

HAUSER, Eugen J. B. Dr.chem., Lic.rer.nat.
Dr. Hauser Consultants, Biel
Principal
Kloosweg 51
CH - 2502 Biel
Tel: +41 32 3234706
Fax: +41 32 3234706
E-Mail: hauser.ub@bluewin.ch

Private address
Kloosweg 51
CH - 2502 Biel
Tel: +41 32 3234706
Fax: +41 32 3234706

Fields of interest
Meat and meat products, poultry, water, additives, antioxidants, aroma active compounds, AAS, electrophoresis, GC, HPLC, MS, spectroscopy, analytical chemistry, food composition, management, microbiology, quality, biotechnology, fermentation, hygiene, minimal processing, packaging, MAP, quality assurance, HACCP

KAUFMANN, Anton Chemiker HTL
Kantonales Labor
Leiter Abteilung Fleisch
Postfach
CH - 8030 Zürich
Tel: +41 01 2525654
Fax: +41 01 2624753

Private address
Mühlefluo 14C
CH - 6414 Oberarth
Tel: +41 41 8554471

Fields of interest
Meat and meat products, poultry, wine, antimicrobials, lipids and fatty acids, drugs, chemometrics, HPLC, MS, analytical chemistry

KIEFFER, Felix Dr.ing.chem. (Biochemist)
Novartis Nutrition
Scientific advisor for human nutrition, retired, Consultant
CH - 3176 Neuenegg
Tel: +41 31 3772696
Fax: +41 31 3772500

Private address
Weissensteinstrasse 93
CH - 3007 Bern

Tel: +41 31 3721957
Fax: +41 31 3721957

Fields of interest
Minerals, trace elements, metabolism, nutrition

KLEIN, Bernard Dr., PD
Laboratoire Cantonal Vaudois
Chimiste Cantonal
Les Croisettes
CH - 1066 Epalinges
Tel: +41 21 3164343
Fax: +41 21 3164300
E-Mail: bernard.klein@lc.vd.ch,

Private address
105, Ch. de Montolieu
CH - 1010 Lausanne
Tel: +41 21 6530438
Fax: +41 21 6530438

Fields of interest
Transgenic food, aroma active compounds, toxic trace elements, AAS, GC, HPLC, adulteration, analytical chemistry, regulative issues, quality assurance, HACCP

KOCH, Herbert Dipl.-Chem., Dr.phil.,
Swiss Federal Veterinary Office
Head of Chemistry Section
Schwarzenburgstr. 161
CH - 3003 Bern
Tel: +41 31 3238539
Fax: +41 31 3233813
E-Mail: herbert.koch@bvet.admin.ch

Private address
Erikaweg 6
CH - 3089 Köniz
Tel: +41 31 9717555

Fields of interest
Fish and fish products, meat and meat products, poultry, additives, drugs, hormones, PAHs, PCBs, AAS, GC, HPLC

LÖLIGER, J. Dr.
Nestlé R & D Center Kemptthal
Deputy Director, Prof of Food Science - University of Lausanne
CH - 8310 Kemptthal
Tel: +41 52 3540689
Fax: +41 52 3540714
E-Mail: juerg.Loeliger@rdke.nestle.com

Fields of interest
Oils and fats, antioxidants, lipids and fatty acids, Maillard reaction products, biotechnology, drying

MIKUSCHKA, Gerhard Dr. phil.
Sanaro SA
Vice-Director
Avenue de Savoie 56
CH - 1896 Vouvry
Tel: +41 24 4811825
Fax: +41 24 4811771
E-Mail: gm@klosterfrau.ch

Private address
56, Rue de Lausanne
CH - 1110 Morges
Tel: +41 21 8018447

Fields of interest
Sweeteners, additives, HPLC, sensory analysis, nutrition, quality, regulative issues, quality assurance, HACCP

MLOTKIEWICZ, Jerzy BA, Ph.D.
Firmenich SA
Senior Reaction Flavour Scientist
Route de Jeunes 1
CH - 1211 Geneva 8
Tel: +41 22 7803744
Fax: +41 22 7803334
E-Mail: jurek.mlotkiewicz@firmenich.com

Private address
19 Rue des Caroubiers
CH - 1227 Carouge, Geneva
Tel: +41 22 3000241

Fields of interest
Meat and meat products, poultry, Aroma active compounds, Maillard reaction products, GC, HPLC, MS, chemical reactions, thermal processing

---

MOSER, Ulrich Dr.
Roche Vitamins Europe Ltd.
Scientist
Postfach 3255
Postfach 3255
CH - 4002 Basel
Tel: +41 61 6882838
Fax: +41 61 6883589
E-Mail: ulrich.moser@roche.com

Private address
Holbeinstr. 85
CH - 4051 Basel
Tel: +41 61 2816606

Fields of interest
Antioxidants, carotenoids, lipids and fatty acids, vitamins, metabolism, nutrition

---

SCHLATTER, Christian Dr.med., Dr.phil. II
Professor for Toxicology, emeritus
E-Mail: schlatter@toxi.biol.ethz.ch

Private address
Peteracher 11
CH - 8126 Zumikon
Tel: +41 1 9180292
Fax: +41 1 9180292

Fields of interest
Alkaloids, biogenic amines, drugs, heterocyclic aromatic amines, hormones, marine toxins, mycotoxins, PAHs, PCBs, pesticides, plant toxins, toxic trace elements

---

SCHMITT, Rudolf Dr.
University of Apllied Science
Professor of Food Microbiology
Rte. du Rawyl 47
CH - 1950 Sion
Tel: +41 27 3243111
Fax: +41 27 3243515
E-Mail: rudolf.schmitt@eiv.ch

Private address
Coin
CH - 1974 Arbaz
Tel: +41 27 3984844

Fields of interest
Microbiology, fermentation, hygiene, quality assurance, HACCP, thermal processing

---

STUDER, Alfred Food Chemist
Nestlé Research Centre
Project Leader
CH - 1000 Lausanne 26
Tel: +41 21 7858036
Fax: +41 21 7858556

Private address
7, Ch. de la Résidence
CH - 1009 Pully
Tel: +41 21 7290523
Fax: +41 21 7290523

Fields of interest
Coffee, tea, cocoa, cosmetics, oils and fats, starch, sugar, honey, sugar alcohols, water, additives, biopolymers, carbohydrates, colours, fat replacers, lipids and fatty acids, polyphenols, HPLC, thermal anlaysis, TLC, analytical chemistry, chemical reactions, food composition, management, drying, fouling and cleaning, freezing, rheology, storage, thermal processing

VAN DAEL, Peter Doctor in Chemistry
Nestlé Research Centre
Project Leader
P.O. Box 44, Vers-chez-les-Blanc
CH - 1000 Lausanne 26
Tel: +41 21 7858184
Fax: +41 21 7858556
E-Mail: peter.van-dael@rdls.nestle.com

Private address
Route du Village
CH - 1614 Granges
Tel: +41 21 9474671

Fields of interest
Dairy, management, nutrition

VISCHER, Michaela Dr.
Vinalytik
Head of Institute
Franzosenstr. 14
CH - 6423 Seewen
Tel: 0041 41 8193468
Fax: 0041 41 8193474
E-Mail: vinalytik@mythen.ch

Fields of interest
Wine, additives, antioxidants, carbohydrates, ethanol, pesticides, GC, HPLC, sensory analysis, food composition, quality, quality assurance, HACCP

VON WIETERSHEIM, Eugen Dr., Ing. Chem.
ETH - Zürich, Retired
Consultant

Private address
Kalchackerstr. 55A
CH - 3047 Bremgarten
Tel: +41 31 3022174

Fields of interest
Beverages, oxidation reactions, toxic trace elements, sensory analysis, sensor technology, quality, drying, packaging, MAP, quality assurance, HACCP

ZOLLER, Otmar Dr.Sc.nat.
Swiss Federal Office of Public Health
Head of Laboratory
CH - 3003 Bern
Tel: +41 31 3229551
Fax: +41 31 3229574
E-Mail: otmar.zoller@bag.admin.ch

Fields of interest
Trace elements, heterocyclic aromatic amines, mycotoxins, plant toxins, toxic trace elements, GC, HPLC, MS, toxicology, irradiation

Turkey

BASMAN, Arzu B.Sc., M.Sc.
Hacettepe University - Ankara
Food Engineering Department
Research Assistant, Teaching Assistant
TR - 06532 Ankara Beytepe
Tel: +90 312 2977100
Fax: +90 312 2354314, 2992123
E-Mail: basman@eti.cc.hun.edu.tr

Private address
Sehit Mustafa Bas Cad. 27/1 Aydinlikevler
TR - 06130 Ankara
Tel: +90 312 3171456

Fields of interest
Bread and cereals, legumes, additives, amino acids, dietary fibres, proteins, electrophoresis, enzymology, food composition, irradiation, rheology

---

CELIK, Süeda Assoc. Prof., Dr.
Hacettepe University - Ankara
Food Engineering Department
TR - 06532 Ankara
Tel: +90 312 2992123
Fax: +90 312 2992123
E-Mail: sueda@eti.cc.hun.edu.tr

Private address
86. Sok. No:16/22 Emek
TR - 06532 Ankara
Tel: +90 312 2238443

Fields of interest
Coffee, tea, cocoa, legumes, additives, amino acids, proteins, electrophoresis, HPLC, spectroscopy, food composition, irradiation, rheology

---

KÖKSEL, Hamit B.Sc., M.Sc., Ph.D.
Hacettepe University - Ankara
Food Engineering Department, Faculty of Engineering
Professor
TR - 06532 Ankara
Tel: +90 312 2977107
Fax: +90 312 2354314
E-Mail: koksel@eti.cc.hun.edu.tr

Private address
8. Cadde 85. Sokak No. 11/2 Emek
TR - 06510 Ankara
Tel: +90 312 2211354

Fields of interest
Bread and cereals, legumes, starch, dietary fibres, proteins, electrophoresis, enzymology, food

composition, irradiation, rheology

YALCIN, Erkan B.Sc., M.Sc.
Hacettepe University - Ankara
Food Engineering Department, Faculty of Engineering
Research Assistent
TR - 06532 Beytepe - Ankara
Tel: +90 312 2977100
Fax: +90 312 2354314
E-Mail: eryalcin@hun.edu.tr

Private address
Hülya sk. A-Blok No. 20/20
TR - 06700 Kucukesat - Ankara
Tel: +90 312 4460888

Fields of interest
Bread and cereals, legumes, additives, amino acids, biopolymers, proteins, electrophoresis, spectroscopy, enzymology, biotechnology

United Kingdom

ADAMS, J. Brian B.Sc., M.Sc., Ph.D.
Campden & Chorleywood Food Research Association
Food Research Scientist
Chipping Campden, Glos. GL 55 6LD
Tel: +44 1386 842015
Fax: +44 1386 841306
E-Mail: b.adams@campden.co.uk

Private address
5, Orchard View, Draycott
Moreton-In-Marsh, Gloucestershire, GLS 69 LW
Tel: +44 1386 700374

Fields of interest
Fruits and vegetables, potatoes, additives, colours, lipids and fatty acids, polyphenols, proteins, spectroscopy, enzymology, thermal processing

AGER, Elaine B.Sc.
ADAS
Analytical Chemist
Woodthorne, Wergs Road
WV6 7XB Wolverhampton
Tel: +44 1902 693256
Fax: +44 1902 743602
E-Mail: elaine.ager@adas.co.uk

Fields of interest
Dairy, oils and fats, amino acids, carotenoids, lipids and fatty acids, vitamins, GC, HPLC

AMES, Jennifer B.Sc., Ph.D.
University of Reading
Department of Food Science and Technology
Reader in Food Chemistry
Whiteknights, P.O. Box 226
Reading RG6 6AP
Tel: +44 1189 318730
Fax: +44 1189 310080
E-Mail: j.m.ames@afnovell.reading.ac.uk

Private address
20, Hartsbourne Road
Reading RG6 5PY
Tel: +44 1189 864150

Fields of interest
Coffee, tea, cocoa, potatoes, aroma active compounds, colours, Maillard reaction products, CE, GC, HPLC, MS, analytical chemistry

ANDERSON, Raymond B.Sc., M.Sc., Ph.D., D.Sc.
Marchington Zymoscience
Owner
The Square, High Ridge
ST14 8LH Uttoxeter, Staffs
Tel: +44 1283 820333
Fax: +44 810 0557620
E-Mail: raymond.anderson@marchington.demen.co.uk

Private address
The Square, High Ridge
ST14 8LH Uttoxeter, Staffs
Tel: +44 1283 820333
Fax: +44 810 0557620

ANDREWS, Anthony M.A., D.Phil., D.Sc., C.Chem., FRSC
University of Wales Institute, Cardiff
School of Applied Sciences, Food and Consumer Science Section
Professor of Food Chemistry
Colchester Avenue
Cardiff CF23 9XR
Tel: +44 29 20416448
Fax: +44 29 20416941
E-Mail: aandrews.uwic.ac.uk

Private address
Tudor Cottage, Branley Lane
Bramley, Tadley, Hants., RG26 5AA
Tel: +44 1256 881278

Fields of interest
Dairy, amino acids, proteins, CE, electrophoresis, HPLC, enzymology, food composition, biotechnology, storage

ASHURST, Philip Roy B.Sc., Ph.D.

Ashurst and Associates
Consulting Chemist
Gooses Foot, Kingstone
Hereford HR2 9HY
Tel: +44 1981 251713
Fax: +44 1981 251715
E-Mail: philip@ashurstassoc.demon.co.uk

Private address
The Gables, Kings Caple
Hereford HR1 4UD
Tel: +44 1432 840408
Fax: +44 1432 840517

Fields of interest
Beer, beverages, fruits and vegetables, additives, aroma active compounds, adulteration, management, microbiology, packaging MAP, thermal processing

---

BAIGRIE, Brian B.Sc., Ph.D.
The Lord Zuckerman Research Centre
Head of Flavour and Trace Analysis
Whiteknights, P.O. Box 234
Reading RG6 2LA
Tel: +44 118 986854253
Fax: +44 118 9868932
E-Mail: brian.d.baigrie@rssl.co.uk

Private address
120, Ellis Road
Crowthorne
Tel: +44 1344 771763

Fields of interest
Coffee, tea, cocoa, aroma active compounds, Maillard reaction products, PAHs, PCBs, pesticides, GC, HPLC, MS, analytical chemistry

---

BAINES, David Allan BA, M.Sc., Ph.D.
Baines Food Consultancy LTD
Consultant, Director
22 Elizabeth Close, Thornbury
BS35 2YN, Bristol
Tel: +44 1454 418104
Fax: +44 1454 418104
E-Mail: davebaines@1way.co.uk

Private address
22 Elizabeth Close, Thornbury
BS35 2YN, Bristol
Tel: +44 1454 418052
Fax: +44 1454 418104

Fields of interest
Meat and meat products, poultry, spices, additives, Maillard reaction products, chemical reactions, food composition, quality, quality assurance, HACCP, flavour technology

---

BEDDOWS, Clifford G. Ph.D., F.R.S.C., C.Chem., FIFST, C.Bio
Leeds Metropolitan University
School of Health Science
Professor, Senior Academic
Calverley
L81 3HE, Leeds
Tel: +44 113 2832600
Fax: +44 113 2833124
E-Mail: c.beddows@lmu.ac.uk

Private address
6 Moorland Gardens
Moortown
Tel: +44 113 2388231
Fax: +44 113 2688231

Fields of interest
Beverages, bread and cereals, fish and fish products, fruits and vegetables, oils and fats, spices, wine, additives, antioxidants, carotenoids, colours, lipids and fatty acids, minerals, oxidation reactions, polyphenols, trace elements, vitamins, HPLC, bioavailability, nutrition, biotechnology

---

BELTON, Peter B.Sc., Ph.D.
Institute of Food Research
Manager Food Quality and Materialdivision
Norwich Research Park, Colney
Norwich NR4 7UA
Tel: +44 1603 255000
Fax: +44 1603 507723
E-Mail: peter.belton@bbsrc.ac.uk

Fields of interest
Bread and cereals, water, biopolymers, carbohydrates, proteins, spectroscopy, high pressure technology

---

BHAT, Mahalingeshwara B.Sc., M.Sc., Ph.D.
Institute of Food Research
Research
Norwich Research Park, Colney
Norwich NR4 7UA
Tel: +44 1603 255396
Fax: +44 1603 507723
E-Mail: mahalingeshwara.bhat@bbsrc.ac.uk

Private address
138 Kingswood Avenue, Taversham Thorpe Marriott
NR4 7UA Norwich
Tel: +44 1603 864475

Fields of interest
Bread and cereals, starch, sugar, honey, sugar alcohols, biopolymers, carbohydrates, dietary fibres, electrophoresis, HPLC, enzymology, biotechnology

BIRCH, Gordon G. B.Sc., Ph.D., D.Sc.
The University of Reading
Department of Food Science and Technology
Emeritus Professor
Whiteknights, P.O. Box 226
Reading RG6 6AP
Tel: +44 1189 318705
Fax: +44 1189 310080

Private address
27 Crosby Hill Drive
GU15 3TZ Camberley
Tel: +44 1276 28104

Fields of interest
Sugar, honey, sugar alcohols, sweeteners, water, additives, carbohydrates, sensory analysis

---

BOAST, Martin Grad RSC
LS Materials
Proprietor
P.O. Box 2900
CO10 0XN Sudburry, Silffowl
Tel: +44 1787 311465
Fax: +44 1787 311465
E-Mail: sales@lsmaterials.demon.co.uk

Fields of interest
Oils and fats, spices, aroma active compounds, sensory analysis, adulteration, quality, quality assurance, HACCP, odour assessments

---

BRANCH, Simon B.Sc. Biology/Chemistry, Ph.D.
Analytical Chemistry
RHM Technology
Head of Chemistry Section
Loro Rank Centre, Lincoln Road
HP12 3QR High Wylombe
Tel: +44 1494 428292
Fax: +44 1494 428114
E-Mail: sbranch@rhmtech.co.uk

Fields of interest
Bread and cereals, dietary fibres, trace elements, toxic trace elements, AAS, adulteration, analytical chemistry, food composition, nutrition

---

BRECHANY, Elizabeth Graduate Royal Society of Chemistry
Hannah Research Institute
Analyst
Mauchline Road
KA6 5HL Ayr
Tel: +44 1292 674098
Fax: +44 1292 674008
E-Mail: brechanye@hri.sari.ac.uk

Private address
Dalmilling Road 11

KA8 0PX Ayr
Tel: +44 1292 268363

Fields of interest
Dairy, aroma active compounds, lipids and fatty acids, GC, HPLC, MS, food composition, biotechnology

---

BROWN, Peter Anthony M.Chem.A., B.Sc.
Lincolne Sutton & Wood Ltd.
Public Analyst
70-80, Oak Street
NR3 3AQ, Norwich
Tel: +44 1603 624555
Fax: +44 1603 629981

Private address
1 Gilbert Way, Cringleford
NR4 7RN Norwich
Tel: +44 1603 456235

Fields of interest
Fruits and vegetables, meat and meat products, poultry, additives, mycotoxins, AAS, GC, HPLC, adulteration, analytical chemistry, food composition

---

BUGLASS, Alan J. B.Sc., Ph.D.
Anglia Polytechnic University
Senior Lecturer (Chemistry Department)
East Road
CB1 1PT Cambridge
Tel: +44 1223 3632712170
E-Mail: aj.buglass@anglia.ac.uk

Private address
Pump Lane 12
Hardwick, Cambridge CB3 7QW
Tel: +44 1954 212156

Fields of interest
Fruits and vegetables, wine, aroma active compounds, lipids and fatty acids, pesticides, GC, HPLC, MS, analytical chemistry

---

BUNN, Cheryl B.Sc.
Mastertaste Part of Kerry Foods
Junior Flavourist
Dreycott Mills, Dursley, Glos.
GL11 5NA Cam, Dursley
Tel: +44 1453 543272
Fax: +44 1453 543272
E-Mail: cheryl-bunn@kerry-ingredients.co.uk

Private address
34 Wharfedale
BS35 2DT Thornbury
Tel: +44 1454 419597

CAMPBELL, Duncan J. B.Sc., Dphil. MChemA

West Yorkshire Analytical Services
Public Analyst

Cliff Lane, County Building
WF1 2TN
Tel: +44 1924 291015
Fax: +44 1924 376388
E-Mail: duncan@wyanalysis.demen.co.uk

Fields of interest
Meat and meat products, poultry, spirits, additives, AAS, HPLC, adulteration, analytical chemistry, food composition, quality, regulative issues

CARDER, John B.Sc.
Badger-Catalytic Ltd
Sales Manager - retired
St. Georges Square
New Malden Surrey
E-Mail: michel@sourism.fsnet.co.uk

Private address
33 King George Gardens
PO194LB Chichester
Tel: +44 1243 839913

Fields of interest
Beer, spirits, wine, antioxidants, biotechnology, fermentation

CLARK, Michael MA, Ph.D.
Unilever plc
Principal Technologist
Colworth House, Sharnbrook
MK44 1LQ Bedford
Tel: +44 1234 222717
E-Mail: michael.g.clark@unilever.com

Fields of interest
Coffee, tea, cocoa, fish and fish products, fruits and vegetables, oils and fats, management, rheology

CLUTTON, David Dr., B.Sc., Ph.D.
Bacardi Martini
Technical Consultant
Westbay Rd.
5015 IDT
Tel: +44 1371 870584
Fax: +44 1371 870584
E-Mail: david.clutton@lineone.net

Private address
Lindsell
Dunmow, Essex CM6 3QG
Tel: +44 1371 870584
Fax: +44 1371 870584

Fields of interest
Spirits, wine, GC, sensory analysis, adulteration, analytical chemistry, management, quality, regulative issues, quality assurance, HACCP

COOPER, Julian B.Sc., Ph.D.
British SugarTechnical Centre
R&D Manager
Norwich Research Park, Colney
Norwich NR4 7UA
Tel: +44 1603 730139
Fax: +44 1603 455874
E-Mail: jmcooper@britishsugar.co.uk, julianmcooper@hotmail.com

Private address
3, Fen View
NR19 1LL Dereham Norfolk
Tel: +44 1362 696429

Fields of interest
Starch, sugar, honey, sugar alcohols, sweeteners, carbohydrates, dietary fibres, HPLC, management, nutrition, drying, filtration

COTTE, Virginie B.Sc.
British American Tobacco
Project Scientist
Regents Park Road, Millbrook
S015 8TL Southampton
Tel: +44 1703 793967

Private address
13 Whitelaw Road
S015 8LH Southampton
Tel: +44 1703 787957

Fields of interest
Aroma active compounds, GC, MS, sensory analysis, analytical chemistry, quality assurance, HACCP, thermal processing

COVENEY, Leslie MSc., MRSC., C.Chem. FIFST
Savant Technologies
Proprietor
20 Wyndham Avenue, Cobham, Surrey
KT 11 IAT
Tel: +44 1932 864915
Fax: +44 1932 864915
E-Mail: lescov@savtech.demon.co.uk

Private address
20 Wyndham Avenue, Cobham, Surrey
KT 11 IAT
Tel: +44 1932 589665

Fields of interest
Dairy, analytical chemistry, food composition, quality, quality assurance, HACCP

United Kingdom

CROSSY, Paul B.Sc., M.Sc.
Eurotest
Laboratory Director
Shirley Ave, Vale Road
SL4 5LH Windsor
Tel: +44 1753 867267
Fax: +44 1753 867847
E-Mail: pcrossy@eurotest.u-net.com

Fields of interest
Water, trace elements, PAHs, PCBs, pesticides, AAS, GC, MS, spectroscopy, analytical chemistry

DAVIES, Robert John B.Sc., M.Chem.A
Humber Authorities Scientific Service
Official Food Control Laboratory (Analyser)
High Streeg 184
HU1 1NE Hull
Tel: +44 1482 327847
Fax: +44 1482 223251

Private address
19, Elveley Drive
HU10 7RT Hull
Tel: +44 1482 658166

Fields of interest
Mycotoxins, adulteration, analytical chemistry, food composition, management, regulative issues

DAVIES, Alan Philipp B.Sc., Ph.D.
Unilever PLC
Unit Manager/Tea Science
Sharnbrook
Bedford MK44 1LQ
Tel: +44 1234 222480
Fax: +44 1234 222844
E-Mail: alan.p.davies@unilever.com

Private address
30 Colchester Way
Bedford MK41 8BG

Fields of interest
Beverages, bread and cereals, coffee, tea, cocoa, antioxidants, oxidation reactions, polyphenols, alkaloids, chemical reactions, food composition

DICKINSON, Eric B.Sc., Ph.D., D.Sc.
University of Leeds
Professor of Food Colloids
Procter Department of Food Science
Leeds L52 9JT
Tel: +44 113 2332956
Fax: +44 113 2332958
E-Mail: e.dickinson@leeds.ac.uk

Fields of interest
Dairy, proteins, high pressure technology, rheology, emulsions, gels, colloids, surfaces, foams

DONALD, Athene MA., Ph.D.
University of Cambridge
Cavendish Lobaratory Department of Physics
Professor
Madlington Road
Cambridge CB3 OHE
Tel: +44 1223 837382
Fax: +44 1223 337000
E-Mail: athene.donald@phy.cam.ac.ut

Fields of interest
Bread and cereals, potatoes, starch, spectroscopy, thermal anlaysis, microscopy, quality, rheology, storage

ENNION, Ronald B.Sc., M.Chem. A
Ruddock & Sherratt
Public Analysist
Hoole Lane
Chester CH2 3EG
Tel: +44 1244 401428
Fax: +44 1244 348868
E-Mail: pa@r-and-s.co.uk

Fields of interest
Adulteration, analytical chemistry, food composition, quality, regulative issues

FARMER, Linda B.Sc., M.Sc., Ph.D.
Queens University of Belfast
Department of Agriculture for Northern Ireland
Project Leader
Newforge Lane; Belfast BT9 5PX
Tel: +44 2890 250666
Fax: +44 2890 669551
E-Mail: linda.farmer@dani.gov.uk

Fields of interest
Fish and fish products, meat and meat products, poultry, Maillard reaction products, oxidation reactions, GC, HPLC, MS, sensory analysis, chemical reactions, quality

FAULDS, Craig B.Sc., Mphil., Ph.D.
Institute of Food Research
Norwich Research Park, Colney
Norwich NR4 7UA
Tel: +44 1603 255000
Fax: +44 1603 507723
E-Mail: craig.faulds@bbsrc.ac.uk

Fields of interest
Beer, bread and cereals, fruits and vegetables, polyphenols, HPLC, spectroscopy, enzymology, metabolism, microbiology, biotechnology

FENWICK, Gruffydd Roger B.Sc., Ph.D.
Institute of Food Research
Research Manager
Norwich Research Park, Colney
Norwich NR4 7UA
Tel: +44 1603 2555277
Fax: +44 1603 507723
E-Mail: roger.fenwick@bbsrc.ac.uk

Fields of interest
Fruits and vegetables, legumes, spices, antioxidants, plant toxins, bioavailability, food composition, management, nutrition, phytoprotectants

FILLERY-TRAVIS, Annette B.Sc., Ph.D.
Institute of Food Research
Senior Scientist
Norwich Research Park, Colney
Norwich NR4 7UA
Tel: +44 1603 255375
Fax: +44 1603 507723
E-Mail: annette.fillery@bbsrc.ac.uk

Private address
6 Berners St.
NR3 2JW Norwich

Fields of interest
Dairy, oils and fats, antioxidants, carotenoids, fat replacers, lipids and fatty acids, proteins, bioavailability, nutrition, rheology

FINNEGAN, Derek B.Sc., M.Sc.
WEETABIX
Station Road, Burton Latimar, Kettering
Northants NN15 5JR
Tel: +44 1536 722181
Fax: +44 1536 420334
E-Mail: def@burtonlatimer.weetabix.com.uk

Private address
3 Waterhouse Gardens
Northants NN15 5TU

Fields of interest
Bread and cereals, aroma active compounds, oxidation reactions, pesticides, GC, MS, analytical chemistry, chemical reactions, quality, regulative issues, quality assurance, HACCP

FINNEY, Graham B.Sc., Ph.D.
Masonline LTD.
Managing Director
Myrtle House, Goose Rye Rd., Worpuesdon, Surrey
GU3 3RJ
Tel: +44 1483 233178
Fax: +44 1483 237263
E-Mail: masonline.ltd@internet.com

Fields of interest
Beverages, dairy, fish and fish products, fruits and vegetables, meat and meat products, poultry, oils and fats, management, quality, regulative issues, quality assurance, HACCP

FISHER, Leonard B.A., B.Sc., M.A., M.Sc., Ph.D., F.R.A.C.I., F.R.S.C.
University of Bristol
Honorary Research Fellow (University of Bristol), Consultant: Firmenich, Geneva, Switzerland
Tyndall Ave
Bristol BS8 ITL
Tel: +44 117 9288753
Fax: +44 117 9255624
E-Mail: len.fisher@bristol.ac.uk

Private address
29, Royal Crescent
Bath BA1 2LT
Tel: +44 1225 424341

Fields of interest
Oils and fats, starch, aroma active compounds, biopolymers, proteins, fouling and cleaning, rheology, emulsions

FLOWERDEN, Mary C.Chem, MRSC
Haarmann + Reimer LTD
Quality Manager
Fieldhouse Lane
Marlow, Buckinghamshire
Tel: +44 1635 562082

Fields of interest
Aroma active compounds, sensory analysis, analytical chemistry, management, quality, quality assurance, HACCP

FRAZIER, Peter B.Sc., Ph.D., C.Chem., Prof., FRSC, FIFST
Food Research and Technology
Consultant, Visiting Professor (Univ. of Reading)
5, Scotts Close, Hilton
Huntington, Cambridge PEI8 9PQ
Tel: +44 1480 831259
Fax: +44 1480 831259
E-Mail: peter-frazier@compuserve.com

Private address
5, Scotts Close
Hilton, Huntington, Cambridgeshire PE18 9PQ
Tel: +44 1480 831259
Fax: +44 1480 831259

Fields of interest
Bread and cereals, legumes, spices, starch, biopolymers, carbohydrates, proteins, biotechnology,

rheology, extrusion cooking (texturisation)

FRYER, John B.Sc., Ph.D., B.A.
Open University
Open University
Associate Lecturer
Milton Keynes

Private address
76 Lakeside Drive
Cardiff CF23 6DG
Tel: +44 2920 762891

Fields of interest
Antimicrobials, vitamins, food composition, nutrition, fermentation

GATES, Leonard Michael B.Sc.
Lionel Hitchen (Essential Oils) LTD
Deputy Technical Manager
Gravel Lane, Barton Stacey
Winchester SO21 3RQ
Tel: +44 1962 760815
Fax: +44 1962 760072
E-Mail: lgates@lhn.co.uk

Private address
1 Linton Drive
Andover SP10 3TT

Fields of interest
Spices, aroma active compounds, GC, MS, sensory analysis, analytical chemistry, quality, quality assurance, HACCP

GILBERT, John B.Sc., M.Sc., Ph.D.
CSL, Ministry of Agriculture, Fisheries and Food
Research Director
Sand Hutton
YO 41 1LZ, York
Tel: +44 1904 462424
Fax: +44 1904 462426

Private address
48, Intwood Road
Norwich NR4 6AA
Tel: +44 1603 452128

Fields of interest
Mycotoxins, GC, HPLC, MS, adulteration, analytical chemistry, chemical reactions, regulative issues, packaging, MAP, quality assurance, HACCP

GOODALL, David M. Dr.
University of York
Department of Chemistry
Reader in Chemistry
University of York
YO10 5DD York
Tel: +44 1904 432754
Fax: +44 1904 432516
E-Mail: dmg1@york.ac.uk

Fields of interest
Biopolymers, carbohydrates, CE, electrophoresis, HPLC, MS, spectroscopy, analytical chemistry

GOODMAN, Bernard B.Sc., Ph.D.
Scottisch Crop Research Institute
Principal Research Scientist
SCRI, Invercowrie
Dundee DD2 5DA
Tel: +44 1382 562731, 568532
Fax: +44 1382 562426
E-Mail: bgoodm@scri.sari.ac.uk

Private address
10 Upper Constitution Street
Dundee DD3 6JP
Tel: +44 1382 804668

Fields of interest
Coffee, tea, cocoa, fruits and vegetables, spices, antioxidants, oxidation reactions, trace elements, spectroscopy, chemical reactions, quality, irradiation, ESR spectroscopy, free radicals

GORDON, Michael M.A., D.Phil.
University of Reading
Food Studies Building
University Lecturer
Whiteknights, P.O. Box 226
Reading RG6 6AP
Tel: +44 118 9316723
Fax: +44 118 9310080

Private address
11 Inston Road
RG6 5QH
Tel: +44 118 9868052

Fields of interest
Oils and fats, antioxidants, lipids and fatty acids, Oxidation reactions, polyphenols, GC, HPLC, analytical chemistry, food composition, nutrition

GRAMSHAW, J.W. B.Sc., Ph.D.
The University of Leeds
Procter Department of Food Science
Senior Lecturer in Food Science
Leeds LS2 9ST
Tel: +44 113 2332958
Fax: +44 113 2332982

Fields of interest
Additives, aroma active compounds, colours, Maillard reaction products, polyphenols, contaminants,

electrophoresis, GC, HPLC, MS, sensor technology, spectroscopy, adulteration, regulative issues

GRENBY, Trevor Hilary B.Sc., Ph.D., FRSC. C.Chem.
London University
Reader
GKT, Guy's Hospital
SE1 9RT London
Tel: +44 171 9554292
Fax: +44 171 9554455
E-Mail: trevor.grenby@kcl.ac.uk

Private address
Tel: +44 1727 862101
Fax: +44 1727 862101

Fields of interest
Bread and cereals, starch, sugar, honey, sugar alcohols, sweeteners, AAS, HPLC, spectroscopy, metabolism, nutrition, regulative issues

GUNSTONE, Frank B.Sc., Ph.D., D.Sc.
Scottish Crop Research Institute
Honorary Professor
SCRI, Invercowrie
Dundee DD2 5DA
Tel: +44 1382 562731
Fax: +44 1382 562426
E-Mail: fgunst@scri.sari.ac.uk, fdg1@st-and.ac.uk

Private address
Nether Rumgally
KY15 5SY Copar Fife
Tel: +44 1334 653613

Fields of interest
Oils and fats, transgenic foods, antioxidants, fat replacers, lipids and fatty acids, GC, spectroscopy, nutrition, biotechnology, genetic engineering

HAMILTON, Colin A. B.Sc.
Bacardi UK LTD
Quality Assurance Manager
Westbay Road, Western Docks
SO51 1DT Southampton
Tel: +44 1703 31800, +44 2380 31800
Fax: +44 1703 226147, +44 2380 226147
E-Mail: chamilto@bacardi.com

Private address
26 Carisbrooke Court, Woodley Lane, Romsey
SO51 7JQ
Tel: +44 1794 516738

Fields of interest
Beer, beverages, spirits, wine, quality, fermentation, quality assurance, HACCP

HAMPTON, Ian B.Sc.
Kent Scientific Services
Public Analyst
8 Abbey Wood Rd.
ME19 6YT
Tel: +44 1732 220001
Fax: +44 1732 220006
E-Mail: ian.hampton@kent.gov.uk

Private address
15 The Orpines, Watering Bury

Fields of interest
Dairy, meat and meat products, poultry, spirits, sugar, honey, sugar alcohols, adulteration, regulative issues

HARRIS, Caroline A. M.Sc.
Pesticides Safety Directorate
Head of Chemistry
Mallard House, Kings Pool, 3 Peas Holme Green
Y051 9LH
Tel: +44 1904 455906
E-Mail: c.a.harris@psd.maff.gsi.gov.uk

Fields of interest
Fruits and vegetables, PCBs, pesticides, analytical chemistry, consumer research, metabolism, regulative issues

HENDERSON, Nick C. B.Sc. (HONS)
Marlow Foods Ltd.
European Contract Manufacturing Manager
Station Road
TS9 7AB, Yarm
Tel: +44 1642 717254
Fax: +44 1642 717229
E-Mail: nick.henderson@marlowfoods.com

Private address
Oughton Close, 17
TS15 9SZ, Yarm
Tel: +44 1642 785847

Fields of interest
Legumes, meat free products, poultry, proteins, management, fermenation, meat free frozen and chilled product ranges

HENSHALL, David B.Sc, Ph.D.
Consultant

Private address
5, Coldicotts Close
Chipping Campden
Tel: +44 1386 841239
Fax: +44 1386 841239

United Kingdom

Fields of interest
Fruits and vegetables, legumes, trace elements, toxic trace elements, analytical chemistry, regulative issues, preservation, storage, thermal processing

---

HEY, Michael James B.Sc., M.Sc., Ph.D.
University of Nottingham
School of Chemistry
Senior Lecturer
University Park
NG7 2RD Nottingham
Tel: +44 115 9513459
Fax: +44 115 9513562
E-Mail: michael.hey@nottingham.ac.uk

Private address
Long Acre 69
NG13 8BN Bingham, Nottingham
Tel: +44 1949 837713

Fields of interest
Beverages, oils and fats, water, antioxidants, biopolymers, lipids and fatty acids, spectroscopy

---

HILLS, Brian B.Sc., M.A., D.Phil
Institute of Food Research
Head of NMR Group, Senior Researcher
Norwich Research Park, Colney
Norwich NR4 7UA
Tel: +44 1603 255378
Fax: +44 1603 507723
E-Mail: brian.hills@bbsrc.ac.uk

Private address
5 Taverham Chase, Taverham
NR8 6NZ Norwich
Tel: +44 1603 868669

Fields of interest
Bread and cereals, fruits and vegetables, oils and fats, starch, biopolymers, carbohydrates, sensory analysis, spectroscopy, quality, thermal processing

---

HITCHCOCK, Christopher BA, Ph.D.
BA, Ph.D.
University of Surrey, (visiting) Professor, retired

Private address
2, Chudleigh Close
Bedford, MK40 3AW
Tel: +44 1234 344407

Fields of interest
Oils and fats, lipids and fatty acids, sensor technology, adulteration, analytical chemistry, food composition, immunology, quality, regulative issues, irradiation

---

HODGSON, Ian Dr., B.Tech., Ph.D.
ISP Alginates (UK) Ltd.
Director
Waterfield
KT20 5HQ Tadworth, Surrey
Tel: +44 1737 377021
Fax: +44 1737 377100
E-Mail: ian.hodgson@monsanto.com

Private address
18, Windermere Way
RH2 0LW Reigate
Tel: +44 1737 769514

Fields of interest
Beer, beverages, additives, biopolymers, carbohydrates, dietary fibres, management, nutrition, regulative issues, rheology

---

HOLST, Birgit Ph.D., Diploma in Food Chemistry
Institute of Food Research Norwich
Visiting Scientist
Norwich Research Park, Colney
Norwich NR4 7UA
Tel: +44 1603 255129
Fax: +44 1603 507723
E-Mail: birgit.holst@bbsrc.ac.uk

Private address
18 Hellesdon Mill Lane
NR6 5BA Norwich
Tel: +44 1603 426614
Fax: +44 1603 426614

Fields of interest
Fruits and vegetables, plant toxins, HPLC, MS, analytical chemistry, bioavailability, metabolism, toxicology, minimal processing, glucosinolates and degradation products

---

HORNE, David S. B.Sc., Ph.D.
Hannah Research Institute
Research
Ayr KA6 5HL
Tel: +44 1292 674095
Fax: +44 1292 674008

Private address
26 Coll Gardens, Dreghorn
KAII 4EA/Irvine

Fields of interest
Dairy, biopolymers, minerals, proteins, sensor technology, food composition, quality, rheology, light scattering

---

HOUGH, Leslie D.Sc., Ph.D., F.R.S.C., F.K.C.
University of London
King's College
Emeritus Professor of Chemistry

Private address
20 Newstead Way
SW19 5HR London, Wimbledon
Tel: +44 181 9474021

Fields of interest
Sugar, honey, sugar alcohols, sweeteners, carbohydrates, chemical research

HOWARD, Julie GRSC
Dupont (U.K.) Limited
Food Analyst
Station Road 40
CB1 2UJ Cambridge
Tel: +44 1223 464500202
Fax: +44 1223 35573
E-Mail: julie.howard@gbr.dupont.com

Private address
Ganwick Close 1
CB9 9JX Haverhill
Tel: +44 1440 707761

Fields of interest
Bread and cereals, legumes, starch, biopolymers, carbohydrates, dietary fibres, GC, HPLC, MS, analytical chemistry

HOWICK, Chris B.Sc., Ph.D.
European Vinyls Corporation
Regulatory Affairs Manager
PO Box 8, The Meath
WA7 4QD Runcorn
Tel: +44 1928 513123
Fax: +44 1928 580539
E-Mail: chris-howick@evc-int.com

Fields of interest
Toxic trace elements, regulative issues, toxicology, packaging, MAP, plastics

IBE, Frank B.Sc., M.Sc., Ph.D., Cchem., MRSC
Central Science Laboratory (MAFF)
YD41 1LZ Sand Hutton, York
Tel: +44 1904 4620002553
Fax: +44 1904 462111
E-Mail: f.ibe@csl.gov.uk

Private address
410 Aylsham Road
NR3 2SA Norwich
Tel: +44 1603 418529

Fields of interest
Meat and meat products, poultry, drugs, hormones, mycotoxins, pesticides, GC, HPLC, Ms, analytical chemistry, irradiation

JONES, Arthur David B.Sc. (HONS) Chemistry
Unilever Research, Colworth Laboratory
Group Leader, Separation Science
Colworth House, Sharnbrook
MK44 1LQ Bedford
Tel: 01234 222563
Fax: 1234 222000
E-Mail: david.ad.jones@unilever.com

Private address
3, Seagle Gardens
Bedford, Bedfordshire MK41 7FE
Tel: +44 1234 364315

Fields of interest
Beverages, dairy, fish and fish products, antioxidants, carotenoids, polyphenols, proteins, heterocyclic aromatic amines, electrophoresis, HPLC, analytical chemistry, management, high pressure technology

JONES, James B.Sc.
Armaghdown Creameries LTD.
Quality Assurance Manager
Unit 30, Greenbank Industrial Estate
BT34 2SJ Newry, Co.Down
Tel: +44 28302 62224
Fax: +44 28302 62224

Fields of interest
Bread and cereals, dairy, meat and meat products, poultry, potatoes, GC, HPLC, spectroscopy, analytical chemistry, management, microbiology, quality, regulative issues, hygiene, packaging, MAP, quality assurance, HACCP

JONES, Alan B.Sc.
John Innes Centre
Applied Genetics Department
Research Scientist
Norwich Research Park, Colney
Norwich NR4 7UA
Tel: +44 1603 452571
Fax: +44 1603 456844
E-Mail: alan.jones@bbsrc.ac.uk

Fields of interest
Legumes, starch, carbohydrates, lipids and fatty acids, GC, HPLC, analytical chemistry, nutrition, biotechnology

KHOKHAR, Santosh Dr.
University of Leeds
Department of Food Science
Senior Lecturer
LS2 9JT
Tel: +44 113 2332975
Fax: +44 113 2332982
E-Mail: s.khokhar@food.leeds.ac.uk

United Kingdom

Fields of interest
Coffee, tea, cocoa, legumes, spices, antioxidants, lipids and fatty acids, minerals, polyphenols, trace elements, vitamins, AAS, HPLC, sensory analysis, analytical chemistry, bioavailability, food composition, nutrition, fermentation, freezing, high pressure technology, thermal processing

KLUPSCH, Robert N.
33 Fairway Avenue
London NW9 0ET
Fields of interest
Beer, wine, additives, food composition, microbiology, fermentation, storage

KROON, Paul B.Sc., Ph.D.
Institute of Food Research
Senior Scientist
Norwich Research Park, Colney
Norwich NR4 7UA
Tel: +44 1603 255236
Fax: +44 1603 507723
E-Mail: paul.kroon@bbsrc.ac.uk

Private address
81 Warwick Street
NR2 3LD Norwich
Tel: +44 1603 611462

Fields of interest
Bread and cereals, fruits and vegetables, antioxidants, polyphenols, proteins, electrophoresis, HPLC, enzymology, metabolism

LALLJIE, Sam B.Sc., M.Sc., Ph.D., EurChem, CChem, MRSC
Unilever Research Colworth
Programme Manager
Colworth House, Sharnbrook
MK44 1LQ Bedford
Tel: +44 1234 222597
Fax: +44 1234 222599
E-Mail: sam.lalljie@unilever.com

Fields of interest
Beverages, cosmetics, fish and fish products, meat and meat products, poultry, additives, amino acids, antimicrobials, antioxidants, colours, proteins, trace elements, alkaloids, drugs, hormones, PAHs, PCBs, pesticides, plant toxins, toxic trace elements, electrophoresis, GC, HPLC, MS, allergology, analytical chemistry, chemical reactions, management, quality, regulative issues, toxicology, biotechnology, hygiene, irradiation, packaging, MAP, preservation, quality assurance, HACCP, storage

LEES, Ronald FRSC, FRSH, MInstFS
Rolton Technical and Scientic Services
Owner
St. Vincent Road, Walton on Thames
KT12 1PB Surrey
Tel: +44 1932 240773
Fax: +44 1932 240773

Private address
St. Vincent Road, Walton on Thames
KT12 1PB Surrey
Tel: +44 1932 240773
Fax: +44 1932 240773

Fields of interest
Coffee, tea, cocoa, sugar, honey, sugar alcohols, lipids and fatty acids, Maillard reaction products, food composition, quality, hygiene, storage

LEIGH, Anthony M.A., CChem., MRSC
Allchem International LTD
Director
Broadway House, 21 Broadway
Maidenhead, Berks SL6 1NJ
Tel: +44 1753 678687
Fax: +44 1753 678689
E-Mail: amleigh@allchem.co.uk

Fields of interest
Beverages, bread and cereals, dairy, sensory analysis, food composition, nutrition

LOHMAN, Joost A. B. Dr.
Bruker UK Limited
Senior R&D Scientist
Banner Lane
CV4 9GH, Coventry
Tel: +44 24 76855200
Fax: +44 24 76465317
E-Mail: joost.lohman@bruker.co.uk

Fields of interest
NMR, instrumentation

LONG, Alan Ph.D., B.Sc., AIC, FRSH
Vega Research
Vega Research
Hon Research Advisor
14 Woodland Rise
Greenford UB6 ORD
Tel: +44 20 89020073
Fax: +44 20 89020073

Private address
Woodland Rise 14
UB6 ORD UK Greenford
Tel: +44 20 89020073
Fax: +44 20 89020073

Fields of interest
Minerals, polyphenols, proteins, trace elements,

vitamins, bioavailability, consumer research, enzymology, food composition, nutrition

---

MAC DOWALL, James C.Chem, MRSC, AIWSc.
Norit LTD
Chief Chemist
Clydesmill Place, Cambuslang Indest
Glasgow G32 8RF
Tel: +44 141 6418841
Fax: +44 141 6410742
E-Mail: macdowall.jim.uk@norit.com

Private address
343 Kelvindale Rd.
GI2 0QU
Tel: +44 141 5167742

Fields of interest
Beer, starch, sugar, honey, sugar alcohols, sweeteners, water, wine, AAS, quality, filtration, adsorption

---

MAKRIS, Dimitris B.Sc., Dip.
Wye College, University of London
Department of Biological Sciences
PhD Student
TN25 5AH, Ashford
Tel: +44 1233 812401475
E-Mail: d.maktis@wye.ac.uk

Fields of interest
Wine, oxidation reactions, polyphenols, HPLC, spectroscopy, TLC, analytical chemistry, food composition

---

MALBON, Raymond M.A., Ph.D., B.A., C.Chem., MRSC
Regency Mowbray Co. LTD
Food Ingredients Manufacturer
Hixon Industrial Estate, Hixon
Staffordshire ST18 0PY
Tel: +44 1889 270554
Fax: +44 1889 270927
E-Mail: sales@regencymowbray.co.uk

Private address
New House Farm, Stanton, Near Ashbourne
Derbyshire DE6 2DD
Tel: +44 1335 324371
Fax: +44 1335 324035

Fields of interest
Additives, colours, manufacturers of flavourings, colours, fruit preperations, chocolate, emulsifiers, stabilisiers

---

MARTIN, Peter Gerard C.Chem., M.Chem.A., FRSC, FIFST
Q.P. Services
Honey-Consultant of the Industry
Orchard Cottage, Crazies Hill
RG 10 8LU Reading
Tel: +44 1189 402212
Fax: +44 1189 401235
E-Mail: honeysci@aol.com

Private address
Orchard Cottage, Crazies Hill
RG 10 8LU Reading
Tel: +44 1189 402212
Fax: +44 1189 401235

Fields of interest
Sugar, honey, sugar alcohols, adulteration, analytical chemistry, management, quality, regulative issues

---

MILLS, Elizabeth Naomi Clare B.Sc., Ph.D.
Institute of Food Research
Senior Post-Doctoral Researcher
Norwich Research Park, Colney
Norwich NR4 7UA
Tel: +44 1603 255000
Fax: +44 1603 507723
E-Mail: clare.mills@bbsrc.ac.uk

Private address
30, Lindford Drive
NT4 6LR Norwich

Fields of interest
Beer, bread and cereals, dairy, proteins, electrophoresis, HPLC, allergology, immunology, high pressure technology, thermal processing

---

MORGAN, Michael R.A. B.Sc., M.Sc., Ph.D.
University of Leeds
Procter Department of Food Science
Food Research
Procter Department of Food Science
Leeds LS2 9JT
Tel: +44 113 2332966
Fax: +44 113 2332982
E-Mail: m.morgan@leeds.ac.uk

Fields of interest
Legumes, hormones, mycotoxins, pesticides, plant toxins, allergology, bioavailability, immunology, metabolism, biotechnology

---

MORRIS, Victor John D.Sc., Ph.D., M.Sc., B.Sc.
Institute of Food Research
Head of Molecular Biophysics Group
Norwich Research Park, Colney
Norwich NR4 7UA
Tel: +44 1603 255271
Fax: +44 1603 507723

Private address
7 Suffield Close
Norwich NR4 6UB
Tel: +44 1603 504338

Fields of interest
Additives, biopolymers, carbohydrates, proteins, rheology, probe microscopy

---

MOTTRAM, Donald B.Sc., Ph.D.
The University of Reading
Department of Food Science and Technology
Professor of Food Chemistry
Whiteknights, P.O. Box 226
Reading RG6 6AP
Tel: +44 118 9316519
Fax: +44 118 9310080
E-Mail: d.s.mottram@reading.ac.uk

Fields of interest
Bread and cereals, fruits and vegetables, meat and meat products, poultry, aroma active compounds, Maillard reaction products, GC, MS, analytical chemistry, food composition

---

MURRAY, Brent Stuart B.Sc., Ph.D.
University of Leeds
Procter Department of Food Science
Lecturer
Leeds LS2 9JT
Tel: +44 1132 332962
Fax: +44 1132 332982

Fields of interest
Dairy, oils and fats, biopolymers, lipids and fatty acids, proteins, electrophoresis, quality, rheology

---

NURSTEN, Harry Erwin B.Sc., Ph.D., D.Sc., F.R.S.C., F.I.F.S.T.
The University of Reading
Department of Food Science and Technology
Professor Emeritus
Whiteknights, P.O. Box 226
Reading RG6 6AP
Tel: +44 1189 316725
Fax: +44 1189 310 080
E-Mail: h.e.nursten@afnovell.reading.ac.at

Fields of interest
Bread and cereals, coffee, tee, cocoa, dairy, legumes, Maillard reaction products, polyphenols, CE, GC

O' NEILL, Ian B.Sc., Ph.D.
Microprevention Tests LTD.
Director
5 Red Hill Close
CB2 5JP Great Shelford, Cambridge
Tel: +44 1223 840291

Fax: +44 1223 840291

Private address
5 Red Hill Close
CB2 5JP Great Shelford, Cambridge
Tel: +44 1223 840291
Fax: +44 1223 840291

Fields of interest
Coffee, tea, cocoa, fruits and vegetables, meat and meat products, poultry, antioxidants, Dietay Fibres, oxidation reactions, heterocyclic aromatic amines, plant toxins, Nitroso compounds, toxicology

---

PARKER, Jane B.Sc., Ph.D.
University of Reading
Department of Food Science and Technology
Research Fellow
Whiteknights, P.O. Box 226
Reading RG6 6AP
Tel: +44 118 9875123
Fax: +44 118 9310080
E-Mail: j.k.parker@afnovell.reading.ac.uk

Private address
43 Deacon Close
RG40 1WF Berkshire
Tel: +44 118 9781524

Fields of interest
Bread and cereals, meat and meat products, poultry, aroma active compounds, Maillard reaction products, GC, MS, sensory analysis, analytical chemistry, chemical reaction

---

PARR, Adrian James B.A., D.phil.
Institute of Food Research
Senior Research Scientist
Norwich Research Park, Colney
Norwich NR4 7UA
Tel: +44 1603 255000
Fax: +44 1603 507723
E-Mail: adrian.parr@bbsrc.ac.uk

Private address
10, Orchard Way, Barrow
Bury St. Edmunds IP29 5BX
Tel: +44 1284 810465

Fields of interest
Fruits and vegetables, transgenic foods, polyphenols, plant toxins, HPLC, spectroscopy, analytical chemistry, food composition, quality, biotechnology

---

PAYNE, Nigel Kenneth B.Sc., M.Sc., MChemA
Pattinson Scientific Services
Public Analyst
Scott House, Penn Street, Scotswood Industrial Estate
NE4 7BG Newcastle upon Tyne

Tel: +44 191 2261300
Fax: +44 191 2261266
E-Mail: nigel.payne@ukontiu.co.uk

Fields of interest
Adulteration, analytical chemistry, food composition, microbiology, quality, regulative issues

---

PEARCE, Steven B.Sc.
Britannia Natural Products LTD.
Managing Director
5, Woodlands Road, Rougham, Suffolk
IP30 9ND Suffolk

Tel: +44 1359 271461
Fax: +44 1359 271672
E-Mail: steven@bnpl.co.uk

Private address
Cair Paravel, George Lane, Suffolk
CO10 7SB Glemsford
Tel: +44 1787 281943

Fields of interest
Beer, beverages, cosmetics, additives, aroma active compounds, HPLC, enzymology, food composition, biotechnology, high pressure technology, quality assurance, HACCP, flavours

---

PEARCE, J. B.Sc, Ph.D.
The Queens University of Belfast
Department of Food Science
Professor of Food Science and Head of Department
Newforge Lane
Belfast BT9 5PX
Tel: +44 1232 255349
Fax: +44 1232 669551
E-Mail: j.pearce@qub.ac.uk

Private address
91, Kings Road
Belfast BT5 7BU
Tel: +44 1232 797540

Fields of interest
Fruits and vegetables, oils and fats, antioxidants, lipids and fatty acids, polyphenols, enzymology, metabolism, nutrition, minimal processing

---

RIDGWAY, Christopher B.Sc., Ph.D., CChem., MRSC B.Sc., Ph.D., Cchem.,
Central Science Laboratory, Ministry of Agriculture, Fisheries and Food
Chemist
Sand Hutton Y041 1LZ, York
Tel: +44 1904 462000
Fax: +44 1904 462111
E-Mail: c.ridgway@csl.gov.uk

Private address
6, Lindley Street
York YO24 4JF
Tel: +44 1904 781236

Fields of interest
Bread and cereals, aroma active compounds, GC, HPLC, sensor technology, spectroscopy, analytical chemistry, quality, preservation, storage

---

RIDLEY, Brian B.Sc.
The Duckworth Group
Senior Flavourist
MI6 9HJ
Tel: +44 161 8720225
Fax: +44 161 8487331
E-Mail: br@duckworth.co.uk

Private address
Tel: +44 207 2777445
Fax: +44 207 2777445

Fields of interest
Aroma active compounds

ROBINS, Elizabeth Naomi Clare M.Sc., B.Sc., Ph.D.
Institute of Food Research
Head of Biophysics and Mathematics Section
Norwich Research Park, Colney
Norwich NR4 7UA
Tel: +44 1603 255209
Fax: +44 1603 507723
E-Mail: margaret.robins@bbsrc.ac.uk

Private address
3 Munnings Close, Swainsthorpe
NR14 8QE Norwich
Tel: +44 1508 470296

Fields of interest
Rheology, emulsions, ultrasonic technology

ROEDIG-PENMAN, Andrea Ph.D.
Smithkline Beecham
Senior Scientist
Nutritional Healthcare, Royal Forest Factory
GL16 8JB Ldeford
Tel: +44 1594 812931
Fax: +44 1594 812903
E-Mail: andrea.x.roedig-penman@sb.com

Private address
2 Morestall Drive
GL7 1 TF, Glos, Cirencester

Fields of interest
Beverages, antioxidants, carbohydrates, polyphenols, vitamins, GC, HPLC, spectroscopy, TLC, analytical chemistry, food composition, nutrition

RUSSELL, Wendy Roslyn B.Sc., Ph.D.
Rowett Research Institute
Nutritional Chemistry
Greenburn Road
AB21 9SB Aberdeen
Tel: +44 1224 712751
Fax: +44 1224 716687
E-Mail: wrr@rri.sari.ac.uk

Fields of interest
Fruits and vegetables, biopolymers, oxidation reactions, polyphenols, electrochemistry, spectroscopy, chemical reactions, metabolism, nutrition, molecular modelling

---

SALTMARSH, Mike B.A.
Inglehurst Foods LTD
Managing Director
53 Blackberry, Four Marks
Hants GU34 5DF
Tel: +44 1420 563413
Fax: +44 1420 563585
E-Mail: inglehurst.foods@btinternet.com

Private address
53 Blackberry, Four Marks
Hants GU34 5DF
Tel: +44 1420 563413
Fax: +44 1420 563585

Fields of interest
Beverages, coffee, tea, cocoa, water, additives, polyphenols, regulative issues, quality assurance, HACCP

SHACKLETON, Ronald C.Chem.
Ron Shackleton Associates
Filtration and Separation in the Food Industry
39 Park Road, Disley
Stockport SK12 2LX
Tel: +44 663 762372
Fax: +44 663 762372
E-Mail: ron@filtration.freeserve.co.uk

Fields of interest
Beer, dairy, spirits, water, wine, filtration

SHEPPARD, Peter D. B.Sc.
Sussex Pharmaceutical LTD.
Technical Manager
Charlwood Rd.
RM19 2HL East Grinstead, Sussex
Tel: +44 1342 311311
Fax: +44 1342 317816
E-Mail: pete.sheppard@sussexpharm.co.uk

Private address
40 Mailsham Rd., Dolegate
BN26 6NL East Sussex

Tel: +44 1323 484921

Fields of interest
Lipids and fatty acids, minerals, vitamins, GC, HPLC, analytical chemistry, nutrition, regulative issues, quality assurance, HACCP

SHERLOCK, John C. Dr., B.Sc., Ph.D.
Ministry of Agriculture, Fisheries and Food
Head of Agriculture and Food Technology Division
St Christopher House, 80-112 Southwark Street
London SE1 0UD
Tel: +44 207 9211151
Fax: +44 207 9211150
E-Mail: john.sherlock@aftd.maff.gov.uk

Private address
113, Carver Hill Road, High Wycombe, Bucks
HP11 2UQ
Tel: +44 1494 537310

Fields of interest
Dairy, fruits and vegetables, legumes, meat and meat products, poultry, potatoes, transgenic food, pesticides, plant toxins, biotechnology

---

SIME, John B.Sc., Ph.D.
Zylepsis Ltd.
Director Research and Development
6 Highpoint, henwood Business Estate
TN24 8DH, Ashford
Tel: +44 1233 660555
Fax: +44 1233 660777
E-Mail: j.sime@zylepsis.co.uk

Private address
31a Canon Woods Way
TN24 9QY Ashford
Tel: +44 468 104328

Fields of interest
Beverages, coffee, tea, cocoa, cosmetics, antimicrobials, antioxidants, aroma active compounds, polyphenols, enzymology, biotechnology

---

SMITH, Linda Bernhardine Margaret Ph.D., B.Sc., MRSC
MI-SI Limited
Technical Writer
211 Oakdene Road, Petts Wood, Kent
BR5 2AR Petts Wood, Kent
Tel: +44 1689 810004
E-Mail: lbmsmith@hotmail.com

Private address
211 Oakdene Road
BR5 2AR Petts Wood, Kent
Tel: +44 1689 810004

Fields of interest
Beverages, coffee, tea, cocoa, fruits and vegetables, aroma active compounds, polyphenols, GC, HPLC, sensory analysis, sensor technology, quality

---

SPIRO, Michael M.Sc., D.Phil., D.Sc., FRSC
Imperial College of Science, Technology and Medicine
Emeritus Professor
London SW7 2AY
Tel: +44 207 5945724
Fax: +44 207 5945801
Fields of interest
Beverages, coffee, tea, cocoa, spices, water, minerals, oxidation reactions, polyphenols, chemical reactions, food composition, extraction kinetics

---

SPOONER, Martin John Richard B.Sc.
Sensory Services Ltd.
Sensory Analysis Consultancy
17 South Drive
Wokingham, Berkshire
Tel: +44 118 9793379
Fax: +44 118 9793379
E-Mail: martinspooner@compuserv.com

Private address
17 South Drive
Wokingham, Berkshire
Tel: +44 118 9771872

Fields of interest
Sensory analysis

STEVENS, Roger Ph.D., D.Sc.
Founding Editor of 'Flavour and Fragrance Journal'
Private address
Low House, Threlkeld
Keswick, Cumbria CA12 4 SQ
Tel: +44 17687 79624

Fields of interest
Beer, spices, spirits, aroma active compounds, ethanol, food composition, fermentation

---

TALBOT, Geoff B.Sc.
Loders Croklaan
Senior Application and Technical Service Manager
London Road, Purfleet
RM19 1SD Essex
Tel: +44 1708 684867
Fax: +44 1708 684872
E-Mail: geoff.talbot@unilever.com

Fields of interest
Oils and fats, lipids and fatty acids, management

TENNANT, David B.Sc., Ph.D.
David Tennant
14 St. Mary's Square
Brighton BN2 1FZ
Tel: +44 1273 241753
Fax: +44 1273 276358
E-Mail: david-t@dircon.co.uk

Private address
14 St. Mary's Square
Brighton BN2 1FZ
Tel: +44 1273 241753
Fax: +44 1273 276358

Fields of interest
Additives, pesticides, regulative issues, biotechnology, novel foods, risk assessment

---

THOMAS, Mark Andrew Llewellyn HWC Chemistry, L.R.S.C.
Express Dairies LTD
Quality Assurance Manager
Oldford, Frome
BA11 2NQ Somerset
Tel: +44 1373 465651
Fax: +44 1373 462443
E-Mail: mark-thomas@express-dairies.co.uk

Private address
22 Wood Leason Avenue, Lyppard Hanford
WR4 0EU Worcester
Tel: +44 956 574001

Fields of interest
Dairy, analytical chemistry, management, quality, hygiene, quality assurance, HACCP

THOMPSON, Kenneth Clive Ph.D., B.Sc.
Alcontrol Laboratories
Chief Scientist
Templeborough House, Mill Close
56O 1BZ Rotherham
Tel: +44 1709 841078
Fax: +44 1709 841024
E-Mail: clive.thompson@beldagroup.com

Private address
4 Hawurth Crescent
56O 3BW Rotherham
Tel: +44 1709 518831

Fields of interest
Meat and meat products, poultry, water, trace elements, toxic trace elements, AAS, spectroscopy, analytical chemistry, food composition, microbiology

---

THORNTON, Raymond E. B.Sc., D.Phil, C.Chem, FRSC
Dibden Partners

United Kingdom

Biochemical Consultancy
Tudor Lodge, Applemore Hill, Dibden
Southampton
Tel: +44 2380 842686
Fax: +44 2380 842686

Private address
Tudor Lodge, Applemore Hill, Dibden
Southampton
Tel: +44 2380 842686
Fax: +44 2380 842686

Fields of interest
Coffee, tea, cocoa, fish and fish products, fruits and vegetables, legumes, oils and fats, spices, transgenic foods, wine, additives, antioxidants, carotenoids, oxidation reactions, polyphenols, trace elements, vitamins, alkaloids, mycotoxins, PAHs, PCBs, toxic trace elements, GC, HPLC, PCR, bioavailability, metabolism, nutrition, regulative issues, toxicology, biotechnology, genetic engineering

---

VAREY, Jane Elizabeth B.Sc., Ph.D.
University of Teesside
Senior Lecturer in Food Science & Nutrition
Borough RD, School of Science & Technology
TS1 3BA Middlesborough
Tel: +44 1642 342464
Fax: +44 1642 342401
E-Mail: j.e.varey@teos.ac.uk,
jane@janevarey.freeserve.co.uk

Private address
8, Farm Lane, Ingleby Barwick
TS17 0RB, Stockton on Tees
Tel: +44 1642 750094

Fields of interest
Minerals, trace elements, toxic trace elements, AAS, spectroscopy, thermal anlaysis, analytical chemistry, bioavailability, chemical reactions, nutrition

---

VELLA, Anthony LRSC MBA
Paragone International LTD.
Director
12 Tithebarn Grove, Calcot, Reading
RG31 7YX
Tel: +44 118 9424892
Fax: +44 118 9624077
E-Mail: avella@totalise.co.uk

Private address
12 Tithebarn Grove
RG31 7YX
Tel: +44 118 9424892
Fax: +44 118 9624077

Fields of interest
Beer, bread and cereals, potatoes, sensory analysis, analytical chemistry, consumer research, quality assurance, HACCP

WALDRON, Keith B.Sc., Ph.D.
Institute of Food Research
Food Research
Norwich Research Park, Colney
Norwich NR4 7UA
Tel: +44 1603 255000 5385
Fax: +44 1603 507723
E-Mail: keith.waldron@bbsrc.ac.uk

Fields of interest
Fruits and vegetables, legumes, potatoes, transgenic foods, antioxidants, biopolymers, carbohydrates, dietary fibres, polyphenols, proteins, vitamins, HPLC

WANG, Rui B.Sc., Ph.D.
British Sugar PLC
Production of Sugar
Oundle Road, British Sugar PLC
PE2 9QU Peterborough
Tel: +44 1733 563171
Fax: +44 1733 422969
E-Mail: rwang@britishsugar.co.uk

Private address
7 Kildare Drive
PE3 9TS Petersborough
Tel: +44 1733 264415

Fields of interest
Sugar, honey, sugar alcohols, carbohydrates, PAHs, GC, HPLC, spectroscopy, analytical chemistry, chemical reactions, filtration, process technology

---

WEBB, Colin B.Sc., Ph.D.
UMIST
Director, Centre for Grain Process Engineering
PO Box 88
M60 IOD, Manchester
Tel: +44 161 2004379
Fax: +44 161 2004399
E-Mail: c.webb@umist.ac.uk

Private address
Talbot Road, Holly Bank
SK13 7DP Glossop
Tel: +44 1457 866726
Fields of interest
Beer, bread and cereals, starch, carbohydrates, ethanol, biotechnology, fermentation

---

WHITEHEAD, John A. B.Sc., CChem., MRSc.
Chesam Botanicals
Business Development Manager
Cunningham House, Westfield Lane
HA3 9ED Harrow, Middlesex

Tel: +44 208 9077779
Fax: +44 208 9091053
E-Mail: johnwhitehead@cheshamchemicals.co.uk

Private address
15 The Avenue
SG4 9RJ Hitchin Herts
Tel: +44 1462 432125
Fax: +44 1462 436461

Fields of interest
Beer, beverages, cosmetics, applications of botanical extracts to food, drink and cosmetics

---

WHITEHOUSE, Brian B.Sc., Ph.D.
Brian Whitehouse Associates
Consultant - Food Regulatory Affairs
6 Church Bank, Richmond Road, Bowdon
Cheshire WA14 3NW
Tel: +44 161 9286681
Fax: +44 161 9248544
E-Mail: brian@churchbank.demon.co.uk

Private address
6 Church Bank, Richmond Road, Bowdon
Cheshire WA14 3NW
Tel: +44 161 9286681
Fax: +44 161 924854

Fields of interest
Starch, sugar, honey, sugar alcohols, sweeteners, additives, carbohydrates, mycotoxins, regulative issues, hygiene, quality assurance, HACCP

WIGGINS, Edgar Hugh B.Sc., Chartered Chemist
Voelker Science Ltd.
Consultant, Senior Chemist, retired
Private address
27, Cotton Road
Potters Bar, Hertfordshire EN6 5JT
Tel: +44 1707 642535

Fields of interest
Beer, beverages, fruits and vegetables, water, additives, toxic trace elements, analytical chemistry, quality, regulative issues

---

WILDE, Peter B.Sc. Biophysics
Institute of Food Research
Food Research
Norwich Research Park, Colney
Norwich NR4 7UA
Tel: +44 1603 255000, 255283
Fax: +44 1603 507723
E-Mail: peter.wilde@bbsrc.ac.uk

Private address
15 Church Lane, Sprowston
Norwich NR7 8AY

Tel: +44 1603 418141

Fields of interest
Beer, bread and cereals, dairy, proteins, quality, emulsions, foams

---

WILLIAMS, Mervyn B.Sc., Ph.D.
Rivington Foods LTD
Operations Controller
Ormside Close
WN2 4HR Wigan, Hindley Green
Tel: +44 1942 255959
Fax: +44 1942 255236
E-Mail: rivifoods@aol.com

Private address
70 Lache Lave
CH4 7LS Chester
Tel: +44 1244 679698

Fields of interest
Bread and cereals, wine, GC, HPLC, MS, analytical chemistry, management, quality, hygiene, packaging, MAP, quality assurance, HACCP

WILLIAMS, John Graham Chemistry Dr. phil., B.Sc.
Tech-Pet Ltd
Managing Director
27, Appey Street
LE16 9AA Market Harborough
Tel: +44 1858 434545
Fax: +44 1858 432828
E-Mail: john@tech-pet.demon.co.uk

Private address
Main Street, Hillbrow
LE7 9LF Tilton on the Hill
Tel: +44 116 2597352
Fax: +44 116 2597352

Fields of interest
Additives, amino acids, Maillard reaction products, GC, HPLC, sensory analysis, analytical chemistry, nutrition, thermal processing, pet food

---

WILLIAMS, Peter Ph.D., CCHEM., FRSC, Prof
North East Wales Institute
Higher Education Institute
Plas Coch, Mold Road
Wrexham Lliilaw
Tel: +44 1978 29308
Fax: +44 1978 290008
E-Mail: williamspa@newi.ac.uk

Private address
24, Parc Govsedd, Govsedd
Holywell, Flintshine CH8 8RP

Tel: +44 352 710733

Fields of interest
Cosmetics, starch, biopolymers, carbohydrates, thermal anlaysis, rheology, hydrocolloids, thickeners, gelling agents, polysaccharides

WILLIAMSON, Gary Ph.D.
Institute of Food Research
Basic Food Research
Norwich Research Park, Colney
Norwich NR4 7UA
Tel: +44 1603 255259
Fax: +44 1603 507723
E-Mail: gary.williamson@bbsrc.ac.uk

Fields of interest
Bread and cereals, coffee, tea, cocoa, fruits and vegetables, antioxidants, polyphenols, HPLC, enzymology, food composition, metabolism, nutrition

WILSON, Peter D.G. B.Sc., Ph.D.
Institute of Food Research
Norwich Research Park, Colney
Norwich NR4 7UA
Tel: +44 1603 255203
Fax: +44 1603 507723
E-Mail: peterdg.wilsonN@bbsrc.ac.uk

Fields of interest
Antimicrobials, carotenoids, bioavailability, microbiology, nutrition, fermentation, packaging, MAP, mathematical modelling

WILSON, Philip L.R.S.C.
Calibre Control International LTD
Applications Chemist
Asher Court, Lyncaslte Way, Appleton
WA4 4ST Warrington
Tel: +44 1925 860401
Fax: +44 1925 860402
E-Mail: philcalibreint@compuserv.com

Fields of interest
Bread and cereals, dairy, starch, water, carbohydrates, proteins, analytical chemistry, food composition, quality, rheology

WILSON, Reginald B.Sc.
Institute of Food Research
Head, Foometrology Section
Norwich Research Park, Colney
Norwich NR4 7UA
Tel: +44 1603 255210
Fax: +44 1603 507723

Private address
9, Greenacres, Little Melton
Norwich NR9 3QU
Tel: +44 1603 813334

Fields of interest
Bread and cereals, coffee, tea, cocoa, fruits and vegetables, meat and meat products, poultry, oils and fats, biopolymers, carbohydrates, chemometrics, sensor technology, spectroscopy, adulteration, analytical chemistry, food composition

WOOD, Brian J.B. Ph.D., B.Sc.
University of Strathclyde
Department of Bioscience and Biotechnology
Reader in Applied Microbiology (retired)
George Street
G1 1XW Glasgow
Tel: +44 141 5482085
E-Mail: b.j.b.wood@strath.ac.uk

Private address
6, Spruce Drive
G66 4DJ Lenzie, Scotland
Tel: +44 141 7763281
Fax: +44 141 7763281

Fields of interest
Bread and cereals, lipids and fatty acids, microbiology, biotechnology, fermentation

ZABETAKIS, Ioannis B.Sc., Ph.D.
University of Leeds
Department of Food Science
Lecturer, M.Sc. Course Tutor
LS2 9JT Leeds
Tel: +44 113 2332965
Fax: +44 113 2332982
E-Mail: y.zabetakis@food.leeds.ac.uk

Private address
9, Berkeley Court, Kelso Street
LS2 9PT Leeds
Tel: +44 113 2460565

Fields of interest
Fish and fish products, fruits and vegetables, aroma active compounds, colours, GC, HPLC, analytical chemistry, enzymology, biotechnology, high pressure technology

Yugoslavia

ESTELECKI, Ilona M.Sc.
A.D. Sojaprotein
Industrijska Zona 66

YU - 21220 Becej
Tel: +381 21 815311
Fax: +381 21 812545
E-Mail: ilona@soyaprotein.com

Private address
Svetozara Calenica 4
YU - 21220 Becej
Tel: +381 21 811389
Fax: +381 21 811389

Fields of interest
Bread and cereals, dairy, meat and meat products, poultry, oils and fats, sweeteners, aroma active compounds, dietary fibres, lipids and fatty acids, proteins, rheology

Names

A

Aalbersberg, Willem Y. 67
Accorsi, Carla Alberta 60
Adamantiadou, Sophia 55
Adams, J. Brian 98
Ager, Elaine 98
Ahola, Maarit 29
Ahvenainen, Juha M.I. 29
Alder, Lutz 39
Amaro Pinto, Rui Manuel 85
Amarowicz, Ryszard 75
Ames, Jennifer 98
Anderson, Raymond 98
Anderssen, Valborg 29
Andrews, Anthony 98
Anklam, Elke 60
Applebye, Ulla 29
Aro, Tarja 29
Ashurst, Philip Roy 98
Assimakopoulou, Angelique 55
Aubourg, Santiago 89
Aura, Anna-Marija 29
Aust, Olivier 39

B

Baars, Aalbert Jan 67
Bachman, Stefania 76
Baigrie, Brian 99
Baines, David Allan 99
Balduck, Paul 10
Balling Engelsen, Soren 26
Baltes, Werner 39
Barber, Berta 90
Barber, Salvador 90
Barron, Luis Javier Rodríguez 90
Basman, Arzu 97
Battaglia, Reto 94
Bauer, Friedrich 7
Bauer, Ulrich 11
Bazulic, Davorin 15
Beddows, Clifford 99
Beentjes, Pieter 68
Beernaert, Hedwig 10
Beljaars, Paul 68
Belton, Peter 99
Benda, Vladimir 19
Benedito, Carmen 90
Berardo, Nicola 60
Berger, Ralf Günter 40
Betsche, Thomas 40
Bhat, Mahalingeshwara 99
Birch, Gordon G. 100
Birlouez, Inès 36
Bjergegaard, Charlotte 26
Boast, Martin 100
Böhm, Josef 11

Böhm, Volker 7
Bontempelli, Gino 61
Bontenbal, Edwin 68
Boskou, Dimitrius 55
Bosset, Jacques Olivier 94
Botrè, Claudio 61
Botrè, Francesco 61
Branch, Simon 100
Brathen, Gudmund 75
Brechany, Elizabeth 100
Brockmann, Anneliese 40
Brockmann, Rainer 40
Brown, Peter Anthony 100
Brunn, Hubertus 40
Bubnik, Zdenek 19
Budde, Jürgen 40
Buglass, Alan J. 100
Büning-Pfaue, Hans 41
Bunn, Cheryl 100
Burini, Giovanni 61
Bykowski, Piotr Jan 76
Byrne, Briege Eileen 59

C

Calvo, Marta Maria 90
Calzolari, Claudio 61
Campanella, Luigi 61
Campbell, Duncan J. 100
Carder, John 101
Catalá, Ramon 90
Cejpek, Karel 20
Celik, Süeda 97
Cerkvenik, Vesna 88
Cheftel, Jean-Claude 36
Chiavaro, Emma 62
Christoph, Norbert 41
Christophersen, Carsten 26
Chroneos, Ioannis 55
Chrysafidis, Dimitrios 55
Chrzanowska, Jozefa 76
Chumchalová, Jana 20
Clark, David 68
Clark, Michael 101
Clutton, David 101
Coimbra, Manuel António 85
Collar Esteve, Concha 91
Cooper, Julian 101
Copíková, Jana 20
Cornelese, Johan 68
Corradini, Claudio 62
Corvi, Claude Albert 94
Cotte, Virginie 101
Coveney, Leslie 101
Crossy, Paul 102
Culik, Jirí 20
Curda, Ladislav 20
Czarnecka, Maria 76
Czarnecki, Zbgniew 76

D

Dalev, Pencho 13
Damiani, Pietro 62
Daphi-Weber, Juliane 41
Davidek, Jiri 20
Davies, Alan Philipp 102
Davies, Robert John 102
Dawihl, Gerd 41
De Block, Jan 10
De Brabander, Hubert 11
De Jong, Jacob 68
De Kruif, Kees C.G. 68
De Ruiter, Gerhard A. 69
De Ruyck, Hendrik 11
Decaris, Bernard 36
Dechow, Arndt 41
Deelstra, Hendrik 11
Delgadillo, Ivonne 85
Demnerova, Katerina 20
Demopolous, Constantinos A. 55
Dettweiler, Gerd 41
Deweghe, Liane 11
Di Luccia, Aldo 62
Di Natale, Corrado 62
Dickinson, Eric 102
Dietrich, Helmut 41
Dirinck, Patrick 11
Doganoc, Darinka Zdenka 88
Dolezal, Marek 21
Donald, Athene 102
Doncheva, Ivanka 13
Dos Santos Baptista, Bráulio 85
Dossena, Arnaldo 62
Dostálová, Jana 21
Drdak, Milan 21
Ducauze, Christian J. 37
Duran, Luis 91
Dziuba, Jerzy 76
Dzwolak, Waldemar 77

E

Eberle, Mike 42
Economides, Anna 55
Eerikäinen, Tero 29
Ehlers, Dorothea 42
Eisenbrand, Gerhard 42
Ekar, Igor 88
Eklund, Trygve 75
Ellen, Geert 69
Empis, José 85
Enders, Peter W. 42
Ennion, Ronald 102
Erning, Dieter 42
Eshuis, Dolf F. 69
Estelecki, Ilona 116
Evangelisti, Filippo 63

Evidente, Antonio 63

F

Farkas, Joszef 57
Farkas, Pavel 86
Farmer, Linda 102
Faulds, Craig 102
Favretto, Luciano 63
Feinberg, Max 37
Fenwick, Gruffydd Roger 103
Ferrara, Lydia 63
Filip, Vladimir 21
Fillery-Travis, Annette 103
Finnegan, Derek 103
Finney, Graham 103
Fisher, Leonard 103
Flowerden, Mary 103
Franzke, Claus 43
Frazier, Peter 103
Frias, Juana 91
Fryer, John 104

G

Gabrielli Favretto, Luciana 63
Galic, Kata 16
Galluser, Anita 94
Garncarek, Barbara 77
Gates, Leonard Michael 104
Gatti, Gian Carlo 63
Gaucheron, Frédéric 37
Gaudard - De Weck, Daniele 94
Gertz, Christian 43
Gil, Ana Maria Pissarra C. 86
Gilbert, John 104
Gladovic, Natasa 88
Glück, Bernfried 43
Gojmerac, Tihomira 16
Golachowski, Antoni 77
Golja, Viviana 88
Gonzalez-Sanjose, Maria Luisa 91
Goodall, David M. 104
Goodman, Bernard 104
Gordon, Michael 104
Graille, Jean 37
Gramshaw, J.W. 104
Grenby, Trevor H. 105
Groothuis, Dirk G. 69
Guillen, M. D. 91
Gunstone, Frank 105
Györi, Zoltán 57

H

Haffke, Helma 43
Hägg, Margareta 30

Hahn, Harald 43
Hajduk, Ewa 77
Hakala, Mari 30
Häkkinen, Sari 30
Hamilton, Colin A. 105
Hampton, Ian 105
Hanewinkel-Meshkini, Susanne 43
Hardi, Jovica 16
Harris, Caroline A. 105
Hasenay, Damir 16
Hauser, Eugen J. B. 95
Havenaar, Robert 69
Hegedusic, Vesna 16
Heiniö, Raija Liisa 30
Heinonen, Marina 30
Heinzler, Matthias 43
Henderson, Nick C. 105
Henle, Thomas 44
Henshall, David 105
Heredia, Antonia 91
Herman, Lieve 11
Hermus, Rudolph J.J. 69
Herraiz Tomico, Tomas 92
Hey, Hanke 44
Hey, Michael James 106
Hietaniemi, Veli 30
Hills, Brian 106
Hils, Arno K. A. 44
Hitchcock, Christopher 106
Hodgson, Ian 106
Holandová, Katerina 21
Holasova, Marie 21
Holch, Klaus 27
Holmer, Gunhild 27
Holst, Birgit 106
Home, Silja 31
Honikel, Karl Otto 44
Honkavaara, Markku 31
Hood, Ted D.E. 59
Hopia, Anu 31
Horne, David S. 106
Horne-Ekman, Maarit 31
Horváth-Mosonyi, Magda 57
Horvatic, Marija 16
Hough, Leslie 106
Howard, Julie 107
Howick, Chris 107
Hrncirik, Karel 21
Hruskar, Mirjana 17
Huopalahti, Rainer 31

I

Ibe, Frank 107
Iciek, Jan 77
Ilari, Jean-Luc 37
Ilmoja, Kalle 28
Ilsbroux, Ingrid 12
Imbs, Boguslaw 77

Ingr, Ivo 21
Iori, Renato 64
Ivanov, Kalintcho 14

J

Jägerstad, Margaretha 93
Janson-Mundel, Ortrun 44
Jarmoluk, Andrzej 78
Järvenpää, Eila 31
Järvi-Käärinäinen, Irma Terhen 35
Jirovetz, Leopold 7
Jones, Alan 107
Jones, Arthur David 107
Jones, James 107
Jörissen, Urban 44
Juarez, Manuela 92

K

Kadi, Andreas 7
Kalac, Pavel 22
Kallio, Heikki 31
Kan, Kees (C.A.) 69
Kas, Jan 22
Kaufmann, Anton 95
Kedzior, Wladyslaw 78
Kellner, Vladimir 22
Kerkvliet, Jacob 70
Keurulainen, Ritva 32
Khokhar, Santosh 107
Kieffer, Felix 95
Kivistö, Laura 32
Klaarenbeek, Tineke 70
Klapec, Tomislav 17
Klein, Bernard 95
Klein, Erich 45
Klostermeyer, Henning 45
Klupsch, Robert N. 108
Kmiecik, Waldemar Andrzej 78
Knieling, Ralph G. 45
Knowles, Michael Ernest 12
Koch, Herbert 95
Köksel, Hamit 97
Komaitis, Michael 56
Kombal, Ralph 45
Kontominas, Michael 56
Kordic, Jasna 17
Korhonen, Hann 32
Kosicki, Zenon 78
Koskenkorva, Anneli 32
Kostyra, Henryk 78
Kovac, Milan 87
Kovac, Spomenka 17
Kovacs, Elisabeth Teresia 57
Kovatcheva-Apostolova, Elena 14

Kozlowska, Halina 78
Krala, Lucjan 79
Kramer, Jörg 45
Krause, Wolfgang 45
Krkoskova, Bernadetta 87
Kroh, Lothar W. 45
Kroll, Jürgen 46
Kroon, Paul 108
Kroyer, Gerhard Th. 7
Krska, Rudolf 7
Kucera, Jiri 22
Kuusisto, Päivi 32
Kvasnicka, Frantisek 22

L

Laakso, Päivi 32
Lach, Günter 46
Lalljie, Sam 108
Lambert, Michael 46
Lampi, Anna-Maija 32
Lampolahti, Soili 33
Landsiedel, Robert 46
Langendam, Johannes 70
Lapveteläinen, Anja 33
Lasztity, Radomir 57
Lautenbacher,
Lutz, Michael 46
Lauwaars, Margreet 70
Le Botlan, Denis 37
Lees, Ronald 108
Lehtonen, Pekka 33
Leibetseder, Josef 8
Leigh, Anthony 108
Leitner, Erich 8
Leloux, Mirjam 70
Lesniak, Wladyslaw 79
Leszczynski, Waclaw 79
Liddle, Peter 37
Lindhauer, Meinolf G. 46
Lingnert, Hans 93
Lisiewska, Zofia Barbara 79
Lisinska, Grazyna 79
Littmann-Nienstedt, Sigrid 47
Llacer, Dolores 92
Lohman, Joost A. B. 108
Löliger, J. 95
Long, Alan 108
Lorenzen, Kay 47
Lossonczy von Losoncz,
Thomas 70
Luckas, Bernd 47
Ludwig, Eberhard 47
Luf, Wolfgang 8
Lugasi, Andrea 57
Luneia, Roberto 64

M

Määttä, Kaisu 33
Mac Dowall, James 109
Maclean, Wim 71
Maier, Hans Gerhard 47
Mäkinen, Marjukka 33
Makris, Dimitris 109
Malbon, Raymond 109
Malisch, Rainer 47
Malwitz, Dietmar 47
Mandic, Milena L. 17
Marchelli, Rosangela 64
Marin, Michèle 38
Marini, Domenico 64
Marioleas, Panagiotis 56
Martelli, Aldo 64
Martin, Gérard 38
Martinez Anaya,
Antonia M. 92
Martinez-Castro, Isabel 92
Marx, Friedhelm 48
Masková, Eva 23
Matilainen, Katri 33
Matissek, Reinhard 48
Matt, Helmuth 68
Matuszek, Tadeusz 79
Mazzei, Franco 65
Mc Donald, Mark 60
Melchior-Larsen, Lone 27
Melzoch, Karel 23
Merlini, Lucio 65
Meunier, Jean-Claude 38
Mikova, Kamila 23
Mikuschka, Gerhard 96
Mills, Elizabeth Naomi
Clare 109
Mischnick, Petra 48
Miskiewicz, Tadeusz 80
Mlotkiewicz, Jerzy 96
Moilanen, Raija 33
Molnar, Pal 58
Molnar-Perl, Ibolya 58
Moreira Da Silva, Aida
Maria 86
Moret, Ivo 65
Morgan, Michael R.A. 109
Morris, Victor John 109
Mörsel, Jörg-Thomas 48
Mosandl, Armin 48
Moser, Ulrich 96
Mottram, Donald 110
Mraz, Igor 87
Munck, Lars 27
Murkovic, Michael 8
Murray, Brent Stuart 110

N

Nehring, Ulrich P. 48

Neicheva, Anastasia 14
Niemi, Sanna-Maria 33
Nöhle, Ulrich 49
Northolt, Martin 71
Nortvedt, Ragnar 75
Noteborn, Hubert 71
Novakovic, Predrag 17
Nowak, Jacek 80
Nursten, Harry Erwin 110

O

O' Neill, Ian 110
O´Brien, John 38
Obretonov, Tzvetan 14
Ochs, Stefan 49
Ogaard Madsen, Jorgen 27
Olano, Agustin 92
Olieman, Cornelis 71
Ollilainen, Velimatti 34
Ooghe, Wilfried 12
Ortola, Concepcion 92
Österdahl,
Bengt-Göran 93
Otte, Jeanette 27
Otteneder, Herbert 49

P

Palasinski,
Mieczyslaw 80
Palka, Krystyna 80
Palla, Gerardo 65
Panovská, Zdenka 23
Parker, Jane 110
Parr, Adrian James 111
Pastoriza, Laura 93
Pavelka, Jiri 23
Payne, Nigel
Kenneth 111
Pearce, J. 111
Pearce, Steven 111
Peksa, Anna 80
Pertoldi Marletta, Giuliana 65
Petz, Michael 49
Pezacki, Wincenty 80
Pfalzgraf, Andreas 49
Pfannhauser, Werner 8
Pietkiewicz, Jerzy Jan 81
Piironen, Vieno 34
Pilizota, Vlasta 18
Pipek, Petr 23
Pischetsrieder,
Monika 49
Pisciotta, Gennaro 65
Plaga-Lodde, Annette 49
Plocková, Milada 24
Pokorny, Jan 24
Pollmer, Udo 49
Popken, Anne M. 50

123

Popov, Dimitre 14
Pospisil, Jasna 18
Poustka, Jan 24
Pozderovic, Andrija 18
Preuss, Axel 50
Primorac, Ljiljana 18
Prugar, Jaroslav 24
Przysiezna, Ewa 81
Pudil, Frantisek 24

R

Ragotzky, Klaus 50
Rahali, Véronique 38
Rantamäkki, Pirjio 34
Räsänen, Janne 34
Rauch, Pavel 25
Reglero, Guillermo J. 93
Restani, Patrizia 66
Ribarova, Fanny 14
Richter, Timo 34
Ridgway, Christopher 111
Ridley, Brian 111
Rihakova, Zdenka 25
Rihar Jereb, Bernarda 89
Rimkus, Gerhard G. 50
Ristow, Reinhard 50
Rizov, Nicolay 14
Robak, Malgorzata 81
Robins, Elizabeth Naomi Clare 111
Rodziewicz, Anna 81
Roedig-Penman, Andrea 111
Roel, Peter 71
Rogerson, Frank 86
Rohn, Sascha 50
Rollin, Patrick 38
Rombouts, Frank 71
Roos, Yrjö Henrik 60
Roozen, Jacques 72
Rops, Wichard 72
Ross-Petersen, Karl Jakob 28
Rüdt, Ulrich 51
Ruiter, Adriaan 72
Russell, Wendy Roslyn 112
Rutkowski, Antoni 81
Rutledge, Douglas 39
Rychtera, Mojmír 25
Rymowicz, Waldemar 81

S

Saarinen, Niina 34
Sakellariou, Christina 56
Salgò, Andràs 58
Salminen, Seppo 34
Saltmarsh, Mike 112
Salvo, Francesco 66
Sandberg, Ann-Sofie 94

Sapunar-Postruznik, Jasenka 18
Sarpeid, Hans-Jacob 75
Sass-Kiss, Agnes 58
Saukko, Maire 34
Schäfer, Karola 51
Schlatter, Christian 96
Schlegel-Zawadzka, Malgorzata 82
Schlett, Claus 51
Schmid, Erich R. 8
Schmidt, Heinz 51
Schmidt, Stefan 87
Schmitt, Rudolf 96
Schmolck, Werner 51
Schneider, Rüdiger 51
Schreier, Peter 51
Schrenk, Dieter 52
Schreyen, Luc 12
Schulte, Erhard 52
Schutte, Leonard 72
Schwack, Wolfgang 52
Schwarz, Walter 25
Schwenke, Klaus Dieter 52
Scordaki, Alexandra 56
Sebecic, Blazenka 18
Seibel, Wilfried 52
Sereno, Alberto M.C. 86
Sforza, Stefano 66
Shackleton, Ronald 112
Sheppard, Peter D. 112
Sherlock, John C. 112
Siebers, Johannes 52
Siegmund, Barbara 9
Siezen, Roland J. 72
Sikora, Marek 82
Sikorski, Zdzistaw 82
Silanpää, Rauno 35
Sime, John 112
Simko, Peter 87
Simon-Sarkadi, Livia 58
Sinigoj-Gacnik, Ksenija 89
Sinková, Terèzia 87
Skibniewska, Krystyna A. 82
Skibsted, Leif H. 28
Skrabka-Blotnicka, Teresa 82
Skrökki, Leila Anneli 35
Slinde, Erik 75
Slominska, Lucyna 82
Smith, Linda Bernhardine Margaret 112
Smolinska, Teresa 83
Sontag, Gerhard 9
Sorensen, Hilmer 28
Sorensen, Susanne 28
Sorhaug, Terje 75
Sousa, Isabel 86
Speer, Karl 53
Spillane, William 60
Spiro, Michael 113
Spooner, Martin John Richard 113

Stanoeva, Elena 15
Stark, Jacques 72
Steegmans, Monique 12
Steinbuch, Erwin 72
Steiner, Ingrid 9
Steinhart, Hans 53
Stelz, Alice 53
Stempiewicz, Regina 29
Stetina, Jiri 113
Stevens, Roger 113
Storgards, Erna 35
Studer, Alfred 96
Subaric, Drago 19
Suhaj, Milan 88
Suiraková, Eva 25
Surowka, Krzysztof 83
Suutarinen, Marjaana 35
Szilágyi, Szilárd 58
Szoltysek, Katarzyna 83

T

Tahvonen, Raija 35
Talbot, Geoff 113
Talou, Thierry 39
Tateo, Fernando 66
Tauscher, Bernhard 53
Temmerman, Guy 12
Tennant, David 113
Ternes, Waldemar 53
Thier, Hans-Peter 53
Thomas, Mark Andrew Llewellyn 113
Thompson, Kenneth Clive 113
Thornton, Raymond E. 113
Tiefenbacher, Karl 9
Timmermans, Eric 73
Tiusanen, Sirkka 35
Toikkanen, Kari 35
Toivo, Jari 35
Tomasik, Piotr 83
Tömösközi, Sándor 59
Topalova, Ivanka Chrstova 15
Tóth-Markus, Marianna 19
Trojan, Marek 83
Trojanowska, Krystyna 84
Troy, Declan J. 60
Tsanev, Roumen 15
Tuomala-Saramäki, Terhi 36
Tykkyläinen, Paavo 36
Tzia, Constantina 56

U

Uccella, Nicola Antonio 66
Ugarcic-Hardi, Zaneta 19
Uijttenboogaart, Theo 73
Ukmar Maljevac, Damjana 89
Ulberth, Franz 9

V

Vahcic, Nada 19
Vahteristo, Liisa 36
Valentini, Giuseppa 66
Valentová, Helena 25
Van Boekel, Tiny 73
Van Dael, Peter 97
Van Den Bosch, Gerrit 73
Van Den Broek, Ad 73
Van Der Schee, Henk A. 73
Van Dokkum, Wim 74
Van Eckert, Renate 9
Van Peteghem, Carlos 12
Van Poppel, Geert A.F.C. 74
Van Renterghem, Roland 13
Van Rhyn, Hans 74
Van Vyncht, Gery 13
Vanluchene, Eric 13
Varadi, Maria 59
Varey, Jane Elizabeth 114
Vedrina-Dragojevic, Irena 19
Velisek, Jan 26
Vella, Anthony 114
Venskutonis, Petras Rimantas 68
Vidal-Valverde, Concepcion 93
Vieths, Stefan 54
Vischer, Michaela 97

Viszkok, Ferenc 59
Vlater, Vladimír 26
Vodrazka, Zdenek 26
Vokk, Raivo 28
Von Rymon Lipinski, Gert-Wolfhard 54
Von Wietersheim, Eugen 97
Voragen, Fons 74

W

Waldron, Keith 114
Wang, Rui 114
Webb, Colin 114
Weder, Jürgen, K.P. 54
Weisz, Richard 10
Whitehead, John A. 114
Whitehouse, Brian 115
Wiedner, Peter 9
Wiggins, Edgar Hugh 115
Wijngaards, Gerrit 74
Wilde, Peter 115
Williams, John Graham 115
Williams, Mervyn 115
Williams, Peter 115
Williamson, Gary 116
Wilska-Jeszka, Jadwiga 84
Wilson, Peter D.G. 116
Wilson, Reginald 120
Winkeler, Heinz-Dieter 54

Wilson, Philip 120
Winkler, Johanna 10
Winterhalter, Peter 54
Witkowska, Danuta 84
Wittenschläger, Lutz 54
Wojtatowicz, Maria 84
Wood, Brian J.B. 116
Wucherpfennig, Karl 54

Y

Yalcin, Erkan 98
Yanishlieva-Maslarova, Nedjalka 15

Z

Zabetakis, Ioannis 116
Zagt, Robert 74
Zegota, Henryk 84
Ziajka, Stefan 84
Zidaric, Metka 89
Ziino, Marisa 66
Zoller, Otmar 97
Zoric, Andreja 89
Zunin, Paola 67
Zyla, Krzysztof 85

Keywords

AAS

Accorsi, Carla Alberta
Beernaert, Hedwig
Beljaars, Paul
Betsche, Thomas
Branch, Simon
Brathen, Gudmund
Brown, Peter Anthony
Burini, Giovanni
Bykowski, Piotr Jan
Campbell, Duncan J.
Corvi, Claude Albert
Crossy, Paul
Deelstra, Hendrik
Dettweiler, Gerd
Deweghe, Liane
Dietrich, Helmut
Doganoc, Darinka Zdenka
Eisenbrand, Gerhard
Ellen, Geert
Erning, Dieter
Favretto, Luciano
Gabrielli Favretto, Luciana
Gaucheron, Frédéric
Glück, Bernfried
Grenby, Trevor H.
Györi, Zoltán
Haffke, Helma
Häkkinen, Sari
Hasenay, Damir
Hauser, Eugen J. B.
Henle, Thomas
Ilmoja, Kalle
Ivanov, Kalintcho
Khokhar, Santosh
Klapec, Tomislav
Klein, Bernard
Kmiecik, Waldemar Andrzej
Koch, Herbert
Lehtonen, Pekka
Lisiewska, Zofia Barbara
Mac Dowall, James
Mandic, Milena L.
Ooghe, Wilfried
Pavelka, Jiri
Pertoldi Marletta, Giuliana
Pfannhauser, Werner
Ristow, Reinhard
Rizov, Nicolay
Sapunar-Postruznik, Jasenka
Schlegel-Zawadzka, Malgorzata
Schlett, Claus
Schneider, Rüdiger
Sebecic, Blazenka
Sinigoj-Gacnik, Ksenija
Steiner, Ingrid
Stelz, Alice
Tahvonen, Raija
Ternes, Waldemar
Thompson, Kenneth Clive
Tiusanen, Sirkka
Trojan, Marek
Van Den Bosch, Gerrit
Van Renterghem, Roland
Varey, Jane Elizabeth
Zyla, Krzysztof

Additives

Adams, J. Brian
Anklam, Elke
Ashurst, Philip Roy
Bachman, Stefania
Baines, David Allan
Balduck, Paul
Barber, Berta
Basman, Arzu
Bauer, Friedrich
Beddows, Clifford G.
Beljaars, Paul
Berger, Ralf Günter
Birch, Gordon G.
Bontenbal, Edwin
Brown, Peter Anthony
Budde, Jürgen
Burini, Giovanni
Campbell, Duncan J.
Celik, Süeda
Chroneos, Ioannis
Chrysafidis, Dimitrios
Collar Esteve, Concha
Davidek, Jiri
Eisenbrand, Gerhard
Ekar, Igor
Eshuis, Dolf F.
Farkas, Pavel
Galluser, Anita
Gatti, Gian Carlo
Glück, Bernfried
Golja, Viviana
Graille, Jean
Gramshaw, J.W.
Hahn, Harald
Hanewinkel-Meshkini, Susanne
Hauser, Eugen J. B.
Hey, Hanke
Hodgson, Ian
Honikel, Karl Otto
Hopia, Anu
Ilari, Jean-Luc
Ilmoja, Kalle
Ivanov, Kalintcho
Jarmoluk, Andrzej
Kadi, Andreas
Kellner, Vladimir
Klupsch, Robert N.
Knieling, Ralph G.
Knowles, Michael Ernest
Koch, Herbert
Koskenkorva, Anneli
Kovac, Milan
Kovacs, Elisabeth Teresia
Kovatcheva-Apostolova, Elena
Krkoskova, Bernadetta
Kroyer, Gerhard Th.
Kucera, Jiri
Kvasnicka, Frantisek
Lalljie, Sam
Lehtonen, Pekka
Lehtonen, Pekka
Lesniak, Wladyslaw
Llacer, Dolores
Llacer, Dolores
Luf, Wolfgang
Malbon, Raymond
Marioleas, Panagiotis
Martin, Gérard
Martinez Anaya, Antonia M.
Mikuschka, Gerhard
Morris, Victor John
Mraz, Igor
Nehring, Ulrich P.
Neicheva, Anastasia
Nöhle, Ulrich
Ortola, Concepcion
Pastoriza, Laura
Pavelka, Jiri
Pearce, Steven
Pezacki, Wincenty
Pietkiewicz, Jerzy Jan
Pipek, Petr
Pollmer, Udo
Ragotzky, Klaus
Restani, Patrizia
Rizov, Nicolay
Robak, Malgorzata
Roel, Peter
Rutkowski, Antoni
Sakellariou, Christina
Salminen, Seppo
Saltmarsh, Mike
Sass-Kiss, Agnes
Saukko, Maire
Scordaki, Alexandra

Seibel, Wilfried
Simko, Peter
Sinigoj-Gacnik, Ksenija
Sinková, Terèzia
Skrabka-Blotnicka, Teresa
Spillane, William
Stanoeva, Elena
Steiner, Ingrid
Studer, Alfred
Suhaj, Milan
Szoltysek, Katarzyna

Tauscher, Bernhard
Tennant, David
Ternes, Waldemar
Thier, Hans-Peter
Thornton, Raymond E.
Tiusanen, Sirkka
Topalova, Ivanka Chrstova
Ugarcic-Hardi, Zaneta
Venskutonis, Petras Rimantas
Vischer, Michaela

Von Rymon Lipinski, Gert-Wolfhard
Whitehouse, Brian
Wiedner, Peter
Wiggins, Edgar Hugh
Williams, John Graham
Wojtatowicz, Maria
Wucherpfennig, Karl
Yalcin, Erkan
Zidaric, Metka
Zoric, Andreja

Adulteration

Anklam, Elke
Ashurst, Philip Roy
Assimakopoulou, Angelique
Battaglia, Reto
Bauer, Friedrich
Beljaars, Paul
Boast, Martin
Boskou, Dimitrius
Branch, Simon
Brown, Peter Anthony
Burini, Giovanni
Calvo, Marta Maria
Campbell, Duncan J.
Christoph, Norbert
Chroneos, Ioannis
Chrysafidis, Dimitrios
Clutton, David
Corradini, Claudio
Damiani, Pietro
Davies, Robert John
De Block, Jan
De Jong, Jacob
Delgadillo, Ivonne
Ennion, Ronald
Farkas, Pavel
Feinberg, Max

Gaucheron, Frédéric
Gertz, Christian
Gilbert, John
Gramshaw, J.W.
Guillen, M. D.
Hampton, Ian
Henle, Thomas
Hitchcock, Christopher
Ilmoja, Kalle
Jirovetz, Leopold
Jörissen, Urban
Juarez, Manuela
Kerkvliet, Jacob
Klein, Bernard
Komaitis, Michael
Kovac, Milan
Kvasnicka, Frantisek
Lambert, Michael
Liddle, Peter
Luf, Wolfgang
Luneia, Roberto
Marini, Domenico
Marioleas, Panagiotis
Martin, Gérard
Martin, Peter Gerard
Mosandl, Armin

Olieman, Cornelis
Ooghe, Wilfried
Otteneder, Herbert
Payne, Nigel Kenneth
Pisciotta, Gennaro
Pollmer, Udo
Pospisil, Jasna
Preuss, Axel
Ristow, Reinhard
Sakellariou, Christina
Sarpeid, Hans-Jacob
Sass-Kiss, Agnes
Schmid, Erich R.
Schreyen, Luc
Scordaki, Alexandra
Skrökki, Leila Anneli
Sontag, Gerhard
Sorensen, Hilmer
Sorensen, Susanne
Suhaj, Milan
Tóth-Markus, Marianna
Ulberth, Franz
Van Eckert, Renate
Van Renterghem, Roland
Van Vyncht, Gery
Wilson, Reginald

Alkaloids

Anklam, Elke
Beljaars, Paul
Betsche, Thomas
Davies, Alan Philipp
Eisenbrand, Gerhard
Empis, José

Evidente, Antonio
Ferrara, Lydia
Herraiz Tomico, Tomas
Ilmoja, Kalle
Lalljie, Sam
Matissek, Reinhard

Noteborn, Hubert
Rollin, Patrick
Schlatter, Christian
Stanoeva, Elena
Tauscher, Bernhard
Thornton, Raymond E.

Allergology

Balduck, Paul
Calvo, Marta Maria
Daphi-Weber, Juliane
Demopolous, Constantinos A.

Henle, Thomas
Klostermeyer, Henning
Kostyra, Henryk
Kucera, Jiri

Lalljie, Sam
Mills, Elizabeth Naomi Clare
Morgan, Michael R.A.
Restani, Patrizia

Sarpeid, Hans-Jacob
Steinhart, Hans

Van Eckert, Renate
Vieths, Stefan

Ziajka, Stefan

Amino acids

Ager, Elaine
Andrews, Anthony
Basman, Arzu
Beljaars, Paul
Berardo, Nicola
Birlouez, Inès
Calzolari, Claudio
Campanella, Luigi
Celik, Süeda
Chiavaro, Emma
Clark, David
Collar Esteve, Concha
Corradini, Claudio
Dalev, Pencho
Di Luccia, Aldo
Dossena, Arnaldo
Dzwolak, Waldemar
Evidente, Antonio
Ferrara, Lydia
Gojmerac, Tihomira
Györi, Zoltán

Hardi, Jovica
Henle, Thomas
Herraiz Tomico, Tomas
Horvatic, Marija
Kovac, Spomenka
Lalljie, Sam
Leibetseder, Josef
Lisiewska, Zofia Barbara
Luckas, Bernd
Ludwig, Eberhard
Luf, Wolfgang
Marchelli, Rosangela
Molnar-Perl, Ibolya
Nortvedt, Ragnar
Nortvedt, Ragnar
Olano, Agustin
Olieman, Cornelis
Ooghe, Wilfried
Pezacki, Wincenty
Przysiezna, Ewa
Ribarova, Fanny

Rizov, Nicolay
Rohn, Sascha
Schäfer, Karola
Sforza, Stefano
Sikora, Marek
Simon-Sarkadi, Livia
Skrabka-Blotnicka, Teresa
Sorensen, Hilmer
Sorensen, Susanne
Szilágyi, Szilárd
Tauscher, Bernhard
Van Boekel, Tiny
Van Den Bosch, Gerrit
Varadi, Maria
Velisek, Jan
Vodrazka, Zdenek
Williams, John Graham
Yalcin, Erkan

Analytical chemistry

Accorsi, Carla Alberta
Adamantiadou, Sophia
Ager, Elaine
Ames, Jennifer
Anklam, Elke
Aust, Olivier
Baigrie, Brian
Battaglia, Reto
Bauer, Friedrich
Beljaars, Paul
Berger, Ralf Günter
Betsche, Thomas
Birlouez, Inès
Bjergegaard, Charlotte
Bontempelli, Gino
Bosset, Jacques Olivier
Botrè, Francesco
Branch, Simon
Brathen, Gudmund
Brown, Peter Anthony
Brunn, Hubertus
Budde, Jürgen
Buglass, Alan J.
Burini, Giovanni
Calvo, Marta Maria
Calzolari, Claudio
Campanella, Luigi
Campbell, Duncan J.
Cejpek, Karel

Cerkvenik, Vesna
Chiavaro, Emma
Christoph, Norbert
Clutton, David
Copíková, Jana
Corradini, Claudio
Cotte, Virginie
Coveney, Leslie
Crossy, Paul
Culik, Jirí
Damiani, Pietro
Davidek, Jiri
Davies, Robert John
Dawihl, Gerd
De Brabander, Hubert
De Jong, Jacob
De Ruiter, Gerhard A.
De Ruyck, Hendrik
Dechow, Arndt
Deelstra, Hendrik
Dettweiler, Gerd
Di Luccia, Aldo
Dietrich, Helmut
Doganoc, Darinka Zdenka
Dolezal, Marek
Doncheva, Ivanka
Dossena, Arnaldo
Dostálová, Jana
Ducauze, Christian J.

Dziuba, Jerzy
Ellen, Geert
Enders, Peter W.
Ennion, Ronald
Erning, Dieter
Evidente, Antonio
Feinberg, Max
Ferrara, Lydia
Finnegan, Derek
Flowerden, Mary
Frias, Juana
Gates, Leonard Michael
Gaucheron, Frédéric
Gertz, Christian
Gilbert, John
Glück, Bernfried
Gojmerac, Tihomira
Golja, Viviana
Goodall, David M.
Gordon, Michael
Graille, Jean
Gramshaw, J.W.
Györi, Zoltán
Haffke, Helma
Hahn, Harald
Hakala, Mari
Häkkinen, Sari
Harris, Caroline A.
Hauser, Eugen J. B.

Heinonen, Marina
Heinzler, Matthias
Henle, Thomas
Henshall, David
Herraiz Tomico, Tomas
Hils, Arno K. A.
Hitchcock, Christopher
Holandová, Katerina
Holasova, Marie
Holst, Birgit
Home, Silja
Howard, Julie
Hrncirik, Karel
Huopalahti, Rainer
Ibe, Frank
Ilmoja, Kalle
Iori, Renato
Ivanov, Kalintcho
Järvenpää, Eila
Jirovetz, Leopold
Jones, Alan
Jones, Arthur David
Jones, James
Jörissen, Urban
Juarez, Manuela
Kaufmann, Anton
Kellner, Vladimir
Khokhar, Santosh
Klein, Bernard
Klein, Erich
Klostermeyer, Henning
Knieling, Ralph G.
Komaitis, Michael
Kontominas, Michael
Kovac, Milan
Kovatcheva-Apostolova, Elena
Kramer, Jörg
Krause, Wolfgang
Kroh, Lothar W.
Kroyer, Gerhard Th.
Krska, Rudolf
Kuusisto, Päivi
Kvasnicka, Frantisek
Laakso, Päivi
Lach, Günter
Lalljie, Sam
Lautenbacher, Lutz-Michael
Le Botlan, Denis
Lehtonen, Pekka
Leitner, Erich
Liddle, Peter
Luckas, Bernd
Luf, Wolfgang
Lugasi, Andrea
Luneia, Roberto
Määttä, Kaisu

Makris, Dimitris
Martin, Gérard
Martin, Peter Gerard
Martinez-Castro, Isabel
Marx, Friedhelm
Mazzei, Franco
Mc Donald, Mark
Melchior Larsen, Lone
Mischnick, Petra
Moilanen, Raija
Molnar-Perl, Ibolya
Moret, Ivo
Mörsel, Jörg-Thomas
Mosandl, Armin
Mottram, Donald
Murkovic, Michael
Nehring, Ulrich P.
Neicheva, Anastasia
Ogaard Madsen, Jorgen
Olano, Agustin
Ollilainen, Velimatti
Ooghe, Wilfried
Österdahl, Bengt-Göran
Otte, Jeanette
Parker, Jane
Parr, Adrian James
Pavelka, Jiri
Payne, Nigel Kenneth
Petz, Michael
Pfalzgraf, Andreas
Pfannhauser, Werner
Piironen, Vieno
Pischetsrieder, Monika
Pisciotta, Gennaro
Popken, Anne M.
Popov, Dimitre
Preuss, Axel
Rahali, Véronique
Rauch, Pavel
Ridgway, Christopher
Rihakova, Zdenka
Rimkus, Gerhard G.
Ristow, Reinhard
Rizov, Nicolay
Roedig-Penman, Andrea
Rollin, Patrick
Rüdt, Ulrich
Rutledge, Douglas
Salgò, András
Sass-Kiss, Agnes
Schäfer, Karola
Schmid, Erich R.
Schmolck, Werner
Schneider, Rüdiger
Schreier, Peter
Schwack, Wolfgang

Sheppard, Peter D.
Siebers, Johannes
Siegmund, Barbara
Simon-Sarkadi, Livia
Sinigoj-Gacnik, Ksenija
Smith, Linda Bernhardine Margaret
Sontag, Gerhard
Sorensen, Hilmer
Sorensen, Susanne
Steegmans, Monique
Steiner, Ingrid
Steinhart, Hans
Studer, Alfred
Suhaj, Milan
Suutarinen, Marjaana
Szilágyi, Szilárd
Talou, Thierry
Tauscher, Bernhard
Ternes, Waldemar
Thier, Hans-Peter
Thomas, Mark Andrew Llewellyn
Thompson, Kenneth Clive
Tiusanen, Sirkka
Toivo, Jari
Tömösközi, Sándor
Topalova, Ivanka Chrstova
Tóth-Markus, Marianna
Tsanev, Roumen
Vahteristo, Liisa
Van Den Bosch, Gerrit
Van Den Broek, Ad
Van Der Schee, Henk A.
Van Eckert, Renate
Van Renterghem, Roland
Van Rhyn, Hans
Van Vyncht, Gery
Varadi, Maria
Varey, Jane Elizabeth
Vella, Anthony
Wang, Rui
Weder, Jürgen, K.P.
Weisz, Richard
Wiedner, Peter
Wiggins, Edgar Hugh
Williams, John Graham
Williams, Mervyn
Wilson, Philip
Wilson, Reginald
Winkeler, Heinz-Dieter
Zabetakis, Ioannis
Zegota, Henryk
Zoric, Andreja

Antimicrobials

Bauer, Ulrich

Bontenbal, Edwin

Cerkvenik, Vesna

Christophersen, Carsten
Chumchalová, Jana
Clark, David
De Jong, Jacob
Decaris, Bernard
Eklund, Trygve
Ellen, Geert
Evidente, Antonio
Farkas, Joszef
Filip, Vladimir
Fryer, John
Groothuis, Dirk G.
Guillen, M. D.

Ilmoja, Kalle
Iori, Renato
Kan, Kees (C.A.)
Kaufmann, Anton
Korhonen, Hann
Lalljie, Sam
Mc Donald, Mark
Northolt, Martin
Nowak, Jacek
Plocková, Milada
Rihakova, Zdenka
Rombouts, Frank
Rychtera, Mojmír

Siezen, Roland J.
Sime, John
Sinigoj-Gacnik, Ksenija
Stark, Jacques
Steiner, Ingrid
Stempiewicz, Regina
Suiraková, Eva
Trojanowska, Krystyna
Van Rhyn, Hans
Van Vyncht, Gery
Vokk, Raivo
Wilson, Peter D.G.

Antioxidants

Amarowicz, Ryszard
Anklam, Elke
Aubourg, Santiago
Aubourg, Santiago
Aura, Anna-Marija
Aust, Olivier
Beddows, Clifford G.
Beljaars, Paul
Berger, Ralf Günter
Birlouez, Inès
Bjergegaard, Charlotte
Böhm, Volker
Bontempelli, Gino
Boskou, Dimitrius
Burini, Giovanni
Campanella, Luigi
Carder, John
Christophersen, Carsten
Corradini, Claudio
Davies, Alan Philipp
Dietrich, Helmut
Dos Santos Baptista, Bráulio
Dossena, Arnaldo
Ekar, Igor
Empis, José
Eshuis, Dolf F.
Fenwick, Gruffydd Roger
Fillery-Travis, Annette
Frias, Juana
Gertz, Christian
Glück, Bernfried
Gonzalez-Sanjose, Maria Luisa
Goodman, Bernard
Gordon, Michael
Graille, Jean
Guillen, M. D.
Gunstone, Frank
Hägg, Margareta
Häkkinen, Sari
Hauser, Eugen J. B.
Heinonen, Marina

Henle, Thomas
Hermus, Rudolph J.J.
Hey, Michael James
Hietaniemi, Veli
Holasova, Marie
Honikel, Karl Otto
Honkavaara, Markku
Hopia, Anu
Ilmoja, Kalle
Iori, Renato
Ivanov, Kalintcho
Jarmoluk, Andrzej
Jones, Arthur David
Khokhar, Santosh
Komaitis, Michael
Korhonen, Hann
Kovac, Milan
Kovatcheva-Apostolova, Elena
Kozlowska, Halina
Kroon, Paul
Kroyer, Gerhard Th.
Kvasnicka, Frantisek
Lalljie, Sam
Lingnert, Hans
Löliger, J.
Luf, Wolfgang
Lugasi, Andrea
Määttä, Kaisu
Mäkinen, Marjukka
Marchelli, Rosangela
Marx, Friedhelm
Melzoch, Karel
Moser, Ulrich
Murkovic, Michael
Neicheva, Anastasia
O' Neill, Ian
Obretonov, Tzvetan
Ochs, Stefan
Pastoriza, Laura
Pearce, J.
Pfannhauser, Werner

Pokorny, Jan
Popken, Anne M.
Pospisil, Jasna
Ragotzky, Klaus
Reglero, Guillermo J.
Ribarova, Fanny
Rizov, Nicolay
Roedig-Penman, Andrea
Rollin, Patrick
Roozen, Jacques
Ross-Petersen, Karl Jakob
Rutkowski, Antoni
Schäfer, Karola
Schmidt, Stefan
Schwarz, Walter
Sime, John
Skibsted, Leif H.
Sorensen, Hilmer
Sorensen, Susanne
Steiner, Ingrid
Steinhart, Hans
Suhaj, Milan
Surowka, Krzysztof
Tauscher, Bernhard
Ternes, Waldemar
Thornton, Raymond E.
Uccella, Nicola Antonio
Van Boekel, Tiny
Van Den Bosch, Gerrit
Vidal-Valverde, Concepcion
Vischer, Michaela
Vokk, Raivo
Waldron, Keith
Williamson, Gary
Wilska-Jeszka, Jadwiga
Winterhalter, Peter
Yanishlieva-Maslarova, Nedjalka
Zegota, Henryk
Zoric, Andreja
Zyla, Krzysztof

Aroma active compounds

Ames, Jennifer
Aro, Tarja
Ashurst, Philip Roy
Baigrie, Brian
Baltes, Werner
Barron, Luis Javier Rodríguez
Berger, Ralf Günter
Boast, Martin
Bosset, Jacques Olivier
Brechany, Elizabeth
Buglass, Alan J.
Christoph, Norbert
Christophersen, Carsten
Coimbra, Manuel António
Cotte, Virginie
Culik, Jirí
Davidek, Jiri
Delgadillo, Ivonne
Dettweiler, Gerd
Dirinck, Patrick
Dolezal, Marek
Dos Santos Baptista, Bráulio
Ducauze, Christian J.
Ehlers, Dorothea
Eisenbrand, Gerhard
Enders, Peter W.
Estelecki, Ilona
Evidente, Antonio
Farkas, Pavel
Favretto, Luciano
Finnegan, Derek
Fisher, Leonard
Flowerden, Mary
Gabrielli Favretto, Luciana
Gates, Leonard Michael

Gramshaw, J.W.
Guillen, M. D.
Hahn, Harald
Hakala, Mari
Hardi, Jovica
Hauser, Eugen J. B.
Hopia, Anu
Huopalahti, Rainer
Ivanov, Kalintcho
Jarmoluk, Andrzej
Jirovetz, Leopold
Kallio, Heikki
Klein, Bernard
Knieling, Ralph G.
Komaitis, Michael
Kovac, Milan
Leitner, Erich
Liddle, Peter
Marin, Michèle
Martinez-Castro, Isabel
Marx, Friedhelm
Melchior Larsen, Lone
Melzoch, Karel
Merlini, Lucio
Mlotkiewicz, Jerzy
Moret, Ivo
Mosandl, Armin
Mottram, Donald
Obretonov, Tzvetan
Parker, Jane
Pearce, Steven
Pezacki, Wincenty
Pokorný, Jan
Popken, Anne M.
Pozderovic, Andrija

Pudil, Frantisek
Reglero, Guillermo J.
Ridgway, Christopher
Ridley, Brian
Rizov, Nicolay
Robak, Malgorzata
Roel, Peter
Rogerson, Frank
Rollin, Patrick
Roozen, Jacques
Schlett, Claus
Schreier, Peter
Schutte, Leonard
Siegmund, Barbara
Sime, John
Smith, Linda Bernhardine Margaret
Stanoeva, Elena
Steinhart, Hans
Stevens, Roger
Talou, Thierry
Tateo, Fernando
Tauscher, Bernhard
Topalova, Ivanka Chrstova
Tóth-Markus, Marianna
Ulberth, Franz
Van Den Bosch, Gerrit
Velisek, Jan
Venskutonis, Petras Rimantas
Vokk, Raivo
Winterhalter, Peter
Witkowska, Danuta
Zabetakis, Ioannis

Beer

Ahvenainen, Juha M.I.
Home, Silja

Ilsbroux, Ingrid
Stevens, Roger

Beverages

Anderssen, Valborg
Anklam, Elke
Ashurst, Philip Roy
Beddows, Clifford G.
Beljaars, Paul
Berger, Ralf Günter
Burini, Giovanni
Calvo, Marta Maria
Christoph, Norbert
Chrysafidis, Dimitrios
Clark, David
Corradini, Claudio

Culik, Jirí
Czarnecki, Zbgniew
Davies, Alan Philipp
De Ruiter, Gerhard A.
Dietrich, Helmut
Dirinck, Patrick
Drdak, Milan
Eberle, Mike
Finney, Graham
Galluser, Anita
Gil, Ana Maria Pissarra C.
Häkkinen, Sari

Hamilton, Colin A.
Hey, Michael James
Hodgson, Ian
Home, Silja
Ilmoja, Kalle
Jones, Arthur David
Kadi, Andreas
Knowles, Michael Ernest
Koskenkorva, Anneli
Lalljie, Sam
Lehtonen, Pekka
Leigh, Anthony

Leitner, Erich
Lesniak, Wladyslaw
Melzoch, Karel
Molnar, Pal
Moret, Ivo
Ooghe, Wilfried
Otteneder, Herbert
Pearce, Steven
Pfannhauser, Werner
Pietkiewicz, Jerzy Jan
Popken, Anne M.
Roedig-Penman, Andrea
Roel, Peter
Roozen, Jacques
Sakellariou, Christina
Saltmarsh, Mike
Sass-Kiss, Agnes
Schreier, Peter
Siegmund, Barbara
Silanpää, Rauno
Sime, John
Skibsted, Leif H.
Smith, Linda Bernhardine Margaret
Sontag, Gerhard
Spiro, Michael
Steiner, Ingrid
Stelz, Alice
Storgards, Erna
Suhaj, Milan
Tóth-Markus, Marianna
Valentová, Helena
Von Wietersheim, Eugen
Whitehead, John A.
Wiedner, Peter
Wiggins, Edgar Hugh
Wilska-Jeszka, Jadwiga
Wucherpfennig, Karl

Bioavailability

Aura, Anna-Marija
Aust, Olivier
Battaglia, Reto
Beddows, Clifford G.
Betsche, Thomas
Böhm, Josef
Böhm, Volker
Bontenbal, Edwin
Clark, David
Damiani, Pietro
Deelstra, Hendrik
Fenwick, Gruffydd Roger
Fillery-Travis, Annette
Frias, Juana
Havenaar, Robert
Henle, Thomas
Hermus, Rudolph J.J.
Holst, Birgit
Horváth-Mosonyi, Magda
Horvatic, Marija
Khokhar, Santosh
Kroyer, Gerhard Th.
Long, Alan
Morgan, Michael R.A.
Murkovic, Michael
Noteborn, Hubert
Pezacki, Wincenty
Rauch, Pavel
Rutkowski, Antoni
Saarinen, Niina
Sandberg, Ann-Sofie
Schlegel-Zawadzka, Malgorzata
Schwack, Wolfgang
Skibniewska, Krystyna A.
Thornton, Raymond E.
Toivo, Jari
Vahteristo, Liisa
Van Den Bosch, Gerrit
Van Dokkum, Wim
Varey, Jane Elizabeth
Vidal-Valverde, Concepcion
Voragen, Fons
Williamson, Gary
Wilson, Peter D.G.
Zagt, Robert
Zyla, Krzysztof

Biogenic amines

Anklam, Elke
Bauer, Friedrich
Beljaars, Paul
Bykowski, Piotr Jan
Dziuba, Jerzy
Glück, Bernfried
Henle, Thomas
Ilmoja, Kalle
Kalac, Pavel
Kellner, Vladimir
Klostermeyer, Henning
Knieling, Ralph G.
Kostyra, Henryk
Kucera, Jiri
Lehtonen, Pekka
Luckas, Bernd
Luf, Wolfgang
Martelli, Aldo
Mazzei, Franco
Northolt, Martin
Pavelka, Jiri
Pezacki, Wincenty
Pfannhauser, Werner
Pipek, Petr
Plocková, Milada
Sass-Kiss, Agnes
Schlatter, Christian
Simon-Sarkadi, Livia
Stanoeva, Elena
Steiner, Ingrid
Surowka, Krzysztof
Szoltysek, Katarzyna
Van Den Bosch, Gerrit
Wiedner, Peter

Biopolymers

Belton, Peter
Betsche, Thomas
Bhat, Mahalingeshwara
Coimbra, Manuel António
Dalev, Pencho
De Kruif, Kees C.G.
Dietrich, Helmut
Dziuba, Jerzy
Eerikäinen, Tero
Evidente, Antonio
Fisher, Leonard
Frazier, Peter
Garncarek, Barbara
Gil, Ana Maria Pissarra C.
Golachowski, Antoni
Goodall, David M.
Henle, Thomas
Hey, Michael James

Hills, Brian
Hodgson, Ian
Horne, David S.
Howard, Julie
Ilari, Jean-Luc
Iori, Renato
Klostermeyer, Henning
Kroll, Jürgen
Kucera, Jiri
Luf, Wolfgang
Mischnick, Petra
Morris, Victor John
Murray, Brent Stuart

Rauch, Pavel
Robak, Malgorzata
Rodziewicz, Anna
Rohn, Sascha
Roos, Yrjö Henrik
Russell, Wendy Roslyn
Salgò, Andràs
Saukko, Maire
Schäfer, Karola
Schwenke, Klaus Dieter
Sforza, Stefano
Studer, Alfred
Tauscher, Bernhard

Tomasik, Piotr
Van Boekel, Tiny
Vodrazka, Zdenek
Voragen, Fons
Waldron, Keith
Wijngaards, Gerrit
Williams, Peter
Williamson, Gary
Wilson, Reginald
Witkowska, Danuta
Yalcin, Erkan
Zegota, Henryk

Biotechnology

Aalbersberg, Willem Y.
Ahvenainen, Juha M.I.
Andrews, Anthony
Balduck, Paul
Barron, Luis Javier Rodríguez
Beddows, Clifford G.
Beljaars, Paul
Berger, Ralf Günter
Betsche, Thomas
Bhat, Mahalingeshwara
Böhm, Josef
Brechany, Elizabeth
Bubnik, Zdenek
Campanella, Luigi
Carder, John
Chumchalová, Jana
Curda, Ladislav
Czarnecki, Zbgniew
Dalev, Pencho
Damiani, Pietro
Dietrich, Helmut
Eerikäinen, Tero
Empis, José
Enders, Peter W.
Evidente, Antonio
Farkas, Pavel
Faulds, Craig
Franzke, Claus
Frazier, Peter
Frias, Juana
Galic, Kata
Graille, Jean
Gunstone, Frank
Hardi, Jovica
Hauser, Eugen J. B.
Henle, Thomas

Heredia, Antonia
Herman, Lieve
Hood, Ted D.E.
Ilsbroux, Ingrid
Iori, Renato
Jones, Alan
Kas, Jan
Klostermeyer, Henning
Knieling, Ralph G.
Koskenkorva, Anneli
Kostyra, Henryk
Kovac, Spomenka
Kozlowska, Halina
Kucera, Jiri
Lalljie, Sam
Lesniak, Wladyslaw
Löliger, J.
Lossonczy von Losoncz, Thomas
Martelli, Aldo
Matilainen, Katri
Melzoch, Karel
Meunier, Jean-Claude
Miskiewicz, Tadeusz
Moilanen, Raija
Moreira Da Silva, Aida Maria
Morgan, Michael R.A.
Noteborn, Hubert
Nowak, Jacek
Otte, Jeanette
Parr, Adrian James
Pearce, Steven
Pezacki, Wincenty
Pietkiewicz, Jerzy Jan
Ragotzky, Klaus
Richter, Timo

Rihakova, Zdenka
Robak, Malgorzata
Rodziewicz, Anna
Roel, Peter
Rohn, Sascha
Rops, Wichard
Rychtera, Mojmír
Rymowicz, Waldemar
Schreier, Peter
Seibel, Wilfried
Sherlock, John C.
Siezen, Roland J.
Sime, John
Sorhaug, Terje
Stark, Jacques
Steinbuch, Erwin
Stempiewicz, Regina
Szoltysek, Katarzyna
Tennant, David
Thornton, Raymond E.
Trojanowska, Krystyna
Van Den Bosch, Gerrit
Vidal-Valverde, Concepcion
Vieths, Stefan
Vodrazka, Zdenek
Voragen, Fons
Webb, Colin
Wijngaards, Gerrit
Witkowska, Danuta
Wojtatowicz, Maria
Wood, Brian J.B.
Yalcin, Erkan
Zabetakis, Ioannis
Zyla, Krzysztof

Bread and cereals

Ahvenainen, Juha M.I.
Alder, Lutz
Amaro Pinto, Rui Manuel

Aura, Anna-Marija
Balling Engelsen, Soren
Barber, Berta

Basman, Arzu
Beddows, Clifford G.
Beljaars, Paul

Belton, Peter
Benedito, Carmen
Berardo, Nicola
Berger, Ralf Günter
Betsche, Thomas
Bhat, Mahalingeshwara
Branch, Simon
Collar Esteve, Concha
Corradini, Claudio
Davies, Alan Philipp
Donald, Athene
Doncheva, Ivanka
Economides, Anna
Estelecki, Ilona
Evidente, Antonio
Faulds, Craig
Finnegan, Derek
Frazier, Peter
Gil, Ana Maria Pissarra C.
Grenby, Trevor H.
Györi, Zoltán
Haffke, Helma
Heiniö, Raija Liisa
Henle, Thomas
Hietaniemi, Veli
Hills, Brian
Horváth-Mosonyi, Magda
Horvatic, Marija

Howard, Julie
Ilsbroux, Ingrid
Jägerstad, Margaretha
Jones, James
Köksel, Hamit
Koskenkorva, Anneli
Kramer, Jörg
Krkoskova, Bernadetta
Kroon, Paul
Lapveteläinen, Anja
Lasztity, Radomir
Leigh, Anthony
Lindhauer, Meinolf G.
Llacer, Dolores
Martinez Anaya, Antonia M.
Melchior Larsen, Lone
Mills, Elizabeth Naomi Clare
Mottram, Donald
Munck, Lars
Niemi, Sanna-Maria
Nursten, Harry Erwin
Ortola, Concepcion
Parker, Jane
Poustka, Jan
Prugar, Jaroslav
Räsänen, Janne
Ridgway, Christopher
Sakellariou, Christina

Salgò, Andràs
Sandberg, Ann-Sofie
Saukko, Maire
Sebecic, Blazenka
Seibel, Wilfried
Siebers, Johannes
Simon-Sarkadi, Livia
Skibniewska, Krystyna A.
Suutarinen, Marjaana
Szilágyi, Szilárd
Szoltysek, Katarzyna
Tiefenbacher, Karl
Tömösközi, Sándor
Ugarcic-Hardi, Zaneta
Van Eckert, Renate
Vella, Anthony
Viszkok, Ferenc
Voragen, Fons
Webb, Colin
Wilde, Peter
Williams, Mervyn
Williamson, Gary
Wilson, Philip
Wilson, Reginald
Wood, Brian J.B.
Yalcin, Erkan
Zyla, Krzysztof

Carbohydrates

Accorsi, Carla Alberta
Amarowicz, Ryszard
Aura, Anna-Marija
Bachman, Stefania
Balling Engelsen, Soren
Baltes, Werner
Beernaert, Hedwig
Beljaars, Paul
Belton, Peter
Benedito, Carmen
Betsche, Thomas
Bhat, Mahalingeshwara
Birch, Gordon G.
Bjergegaard, Charlotte
Bubnik, Zdenek
Calvo, Marta Maria
Chrysafidis, Dimitrios
Coimbra, Manuel António
Cooper, Julian
Copíková, Jana
Corradini, Claudio
De Ruiter, Gerhard A.
Decaris, Bernard
Delgadillo, Ivonne
Deweghe, Liane
Dietrich, Helmut
Dostálová, Jana
Duran, Luis
Enders, Peter W.

Evidente, Antonio
Frazier, Peter
Frias, Juana
Garncarek, Barbara
Gil, Ana Maria Pissarra C.
Golachowski, Antoni
Goodall, David M.
Grenby, Trevor H.
Havenaar, Robert
Henle, Thomas
Heredia, Antonia
Hills, Brian
Hodgson, Ian
Home, Silja
Hough, Leslie
Howard, Julie
Ilari, Jean-Luc
Ivanov, Kalintcho
Jones, Alan
Kosicki, Zenon
Kozlowska, Halina
Krkoskova, Bernadetta
Kroh, Lothar W.
Le Botlan, Denis
Lehtonen, Pekka
Leszczynski, Waclaw
Lindhauer, Meinolf G.
Lisinska, Grazyna
Martinez Anaya, Antonia M.

Martinez-Castro, Isabel
Mischnick, Petra
Moilanen, Raija
Molnar-Perl, Ibolya
Moreira Da Silva, Aida Maria
Morris, Victor John
Olano, Agustin
Olieman, Cornelis
Palasinski, Mieczyslaw
Pastoriza, Laura
Peksa, Anna
Rizov, Nicolay
Rodziewicz, Anna
Roedig-Penman, Andrea
Rollin, Patrick
Rombouts, Frank
Roos, Yrjö Henrik
Salminen, Seppo
Sikora, Marek
Slominska, Lucyna
Sorensen, Hilmer
Sousa, Isabel
Steegmans, Monique
Steinhart, Hans
Studer, Alfred
Tauscher, Bernhard
Thier, Hans-Peter
Tiefenbacher, Karl
Timmermans, Eric

Tomasik, Piotr
Van Den Bosch, Gerrit
Vidal-Valverde, Concepcion
Vischer, Michaela
Voragen, Fons
Waldron, Keith

Wang, Rui
Webb, Colin
Whitehouse, Brian
Williams, Peter
Wilson, Philip
Wilson, Reginald

Witkowska, Danuta
Yanishlieva-Maslarova, Nedjalka
Zegota, Henryk
Zyla, Krzysztof

Carotenoids

Ager, Elaine
Aura, Anna-Marija
Aust, Olivier
Beddows, Clifford G.
Betsche, Thomas
Böhm, Volker
Brockmann, Rainer
Dietrich, Helmut
Empis, José
Fillery-Travis, Annette
Graille, Jean
Hägg, Margareta

Häkkinen, Sari
Heinonen, Marina
Hermus, Rudolph J.J.
Jones, Arthur David
Kmiecik, Waldemar Andrzej
Lindhauer, Meinolf G.
Lisiewska, Zofia Barbara
Moser, Ulrich
Popken, Anne M.
Ribarova, Fanny
Rogerson, Frank
Rychtera, Mojmír

Schwack, Wolfgang
Skibsted, Leif H.
Szoltysek, Katarzyna
Tauscher, Bernhard
Thornton, Raymond E.
Ugarcic-Hardi, Zaneta
Ulberth, Franz
Van Den Bosch, Gerrit
Wilson, Peter D.G.
Winterhalter, Peter

Capillary electrophoresis

Ames, Jennifer
Anderssen, Valborg
Andrews, Anthony
Anklam, Elke
Baines, David Allan
Bazulic, Davorin
Beernaert, Hedwig
Beljaars, Paul
Bjergegaard, Charlotte
Boast, Martin
Calzolari, Claudio
Corradini, Claudio
Cotte, Virginie
Coveney, Leslie
Curda, Ladislav
De Block, Jan
Di Luccia, Aldo
Doncheva, Ivanka
Dossena, Arnaldo
Ducauze, Christian J.
Dzwolak, Waldemar
Economides, Anna
Ellen, Geert
Enders, Peter W.
Erning, Dieter
Filip, Vladimir
Finnegan, Derek
Finney, Graham
Flowerden, Mary
Frias, Juana
Galluser, Anita
Gates, Leonard Michael
Gatti, Gian Carlo
Gilbert, John

Gladovic, Natasa
Glück, Bernfried
Gojmerac, Tihomira
Goodall, David M.
Groothuis, Dirk G.
Hägg, Margareta
Hamilton, Colin A.
Hauser, Eugen J. B.
Henle, Thomas
Herman, Lieve
Hey, Hanke
Hruskar, Mirjana
Huopalahti, Rainer
Ilmoja, Kalle
Janson-Mundel, Ortrun
Jones, James
Kadi, Andreas
Kalac, Pavel
Kedzior, Wladyslaw
Kivistö, Laura
Klaarenbeek, Tineke
Klein, Bernard
Klein, Erich
Knieling, Ralph G.
Koskenkorva, Anneli
Kovac, Milan
Kramer, Jörg
Krause, Wolfgang
Kuusisto, Päivi
Lalljie, Sam
Lambert, Michael
Lorenzen, Kay
Lossonczy Von Losoncz, Thomas

Luckas, Bernd
Luf, Wolfgang
Maclean, Wim
Malwitz, Dietmar
Marchelli, Rosangela
Marini, Domenico
Martelli, Aldo
Melzoch, Karel
Mikova, Kamila
Mikuschka, Gerhard
Mischnick, Petra
Moilanen, Raija
Molnar, Pal
Mosandl, Armin
Mraz, Igor
Nöhle, Ulrich
Nursten, Harry Erwin
Olieman, Cornelis
Otte, Jeanette
Pearce, Steven
Pipek, Petr
Pisciotta, Gennaro
Rizov, Nicolay
Rombouts, Frank
Rutledge, Douglas
Saltmarsh, Mike
Saukko, Maire
Schmitt, Rudolf
Schmolck, Werner
Schreyen, Luc
Schwarz, Walter
Sheppard, Peter D.
Sorensen, Hilmer
Sorensen, Susanne

Steinbuch, Erwin
Steiner, Ingrid
Stetina, Jiri
Storgards, Erna
Suhaj, Milan
Szoltysek, Katarzyna
Thomas, Mark Andrew
Llewellyn
Tiefenbacher, Karl

Ukmar Maljevac, Damjana
Vahcic, Nada
Van Den Bosch, Gerrit
Van Eckert, Renate
Vanluchene, Eric
Varadi, Maria
Varadi, Maria
Vella, Anthony
Venskutonis, Petras Rimantas

Vischer, Michaela
Von Wietersheim, Eugen
Weisz, Richard
Whitehouse, Brian
Williams, Mervyn
Wittenschläger, Lutz
Zagt, Robert

Chemical reactions

Amarowicz, Ryszard
Baines, David Allan
Benedito, Carmen
Berger, Ralf Günter
Bontenbal, Edwin
Collar Esteve, Concha
Damiani, Pietro
Davidek, Jiri
Davies, Alan Philipp
Dolezal, Marek
Dos Santos Baptista, Bráulio
Dossena, Arnaldo
Dostálová, Jana
Dziuba, Jerzy
Eisenbrand, Gerhard
Evidente, Antonio
Farmer, Linda
Filip, Vladimir
Finnegan, Derek
Gaucheron, Frédéric
Gilbert, John
Goodman, Bernard
Graille, Jean

Henle, Thomas
Herraiz Tomico, Tomas
Klostermeyer, Henning
Knowles, Michael Ernest
Kovac, Milan
Krala, Lucjan
Kroh, Lothar W.
Kroll, Jürgen
Lalljie, Sam
Lingnert, Hans
Luf, Wolfgang
Mäkinen, Marjukka
Mischnick, Petra
Mlotkiewicz, Jerzy
Mörsel, Jörg-Thomas
O´Brien, John
Obretonov, Tzvetan
Olano, Agustin
Piironen, Vieno
Pischetsrieder, Monika
Pisciotta, Gennaro
Rogerson, Frank
Rollin, Patrick

Roos, Yrjö Henrik
Roozen, Jacques
Russell, Wendy Roslyn
Schmidt, Stefan
Schreier, Peter
Schwack, Wolfgang
Schwenke, Klaus Dieter
Sforza, Stefano
Sikorski, Zdzistaw
Sorensen, Hilmer
Spiro, Michael
Stanoeva, Elena
Studer, Alfred
Tauscher, Bernhard
Thier, Hans-Peter
Tomasik, Piotr
Van Boekel, Tiny
Varey, Jane Elizabeth
Velisek, Jan
Voragen, Fons
Wang, Rui

Chemometrics

Balling Engelsen, Soren
Berardo, Nicola
Brathen, Gudmund
Büning-Pfaue, Hans
Burini, Giovanni
Curda, Ladislav
Damiani, Pietro
Di Natale, Corrado
Dirinck, Patrick
Ducauze, Christian J.
Eerikäinen, Tero
Enders, Peter W.
Favretto, Luciano
Feinberg, Max
Frias, Juana

Gabrielli Favretto, Luciana
Gertz, Christian
Gil, Ana Maria Pissarra C.
Guillen, M. D.
Ilari, Jean-Luc
Jirovetz, Leopold
Kaufmann, Anton
Knieling, Ralph G.
Luf, Wolfgang
Marini, Domenico
Martin, Gérard
Martinez Anaya, Antonia M.
Moret, Ivo
Munck, Lars
Nortvedt, Ragnar

Noteborn, Hubert
Petz, Michael
Pisciotta, Gennaro
Pudil, Frantisek
Rutledge, Douglas
Salgò, András
Sarpeid, Hans-Jacob
Smith, Linda Bernhardine Margaret
Suutarinen, Marjaana
Trojan, Marek
Ulberth, Franz
Van Den Bosch, Gerrit
Wilson, Reginald

Coffee, tea, cocoa

Amarowicz, Ryszard
Ames, Jennifer
Anklam, Elke
Aura, Anna-Marija
Baigrie, Brian
Baltes, Werner
Beljaars, Paul
Böhm, Volker
Celik, Süeda
Chiavaro, Emma
Chrysafidis, Dimitrios
Clark, Michael
Coimbra, Manuel António
Copíková, Jana
Davies, Alan Philipp
Dirinck, Patrick
Doncheva, Ivanka

Galluser, Anita
Goodman, Bernard
Hanewinkel-Meshkini, Susanne
Henle, Thomas
Jörissen, Urban
Khokhar, Santosh
Kosicki, Zenon
Lees, Ronald
Ludwig, Eberhard
Maier, Hans Gerhard
Martelli, Aldo
Matissek, Reinhard
Nöhle, Ulrich
Nursten, Harry Erwin
O' Neill, Ian
Ochs, Stefan

Panovská, Zdenka
Rops, Wichard
Sakellariou, Christina
Saltmarsh, Mike
Sikora, Marek
Sime, John
Smith, Linda Bernhardine Margaret
Speer, Karl
Spiro, Michael
Steinhart, Hans
Studer, Alfred
Thornton, Raymond E.
Weisz, Richard
Wilson, Reginald
Winterhalter, Peter
Zidaric, Metka

Colour

Adams, J. Brian
Ames, Jennifer
Beddows, Clifford G.
Berger, Ralf Günter
Brockmann, Rainer
Christophersen, Carsten
Chrysafidis, Dimitrios
Drdak, Milan
Ekar, Igor
Glück, Bernfried
Gonzalez-Sanjose, Maria Luisa
Gramshaw, J.W.
Haffke, Helma
Heinonen, Marina

Hopia, Anu
Huopalahti, Rainer
Ilmoja, Kalle
Kovac, Milan
Krala, Lucjan
Kroyer, Gerhard Th.
Lalljie, Sam
Malbon, Raymond
Matissek, Reinhard
Obretonov, Tzvetan
Pezacki, Wincenty
Pilizota, Vlasta
Pipek, Petr
Pospisil, Jasna

Pozderovic, Andrija
Przysiezna, Ewa
Rizov, Nicolay
Simko, Peter
Skrabka-Blotnicka, Teresa
Studer, Alfred
Subaric, Drago
Tauscher, Bernhard
Wilska-Jeszka, Jadwiga
Winkeler, Heinz-Dieter
Winterhalter, Peter
Zabetakis, Ioannis

Consumer research

Ahola, Maarit
Applebye, Ulla
Assimakopoulou, Angelique
Beljaars, Paul
Benda, Vladimir
Bubnik, Zdenek
Calzolari, Claudio
Cerkvenik, Vesna
Dostálová, Jana
Gonzalez-Sanjose, Maria Luisa

Harris, Caroline A.
Heiniö, Raija Liisa
Ilmoja, Kalle
Imbs, Boguslaw
Kadi, Andreas
Kedzior, Wladyslaw
Klein, Erich
Knowles, Michael Ernest
Lautenbacher, Lutz-Michael
Long, Alan

Ooghe, Wilfried
Panovská, Zdenka
Pezacki, Wincenty
Prugar, Jaroslav
Schlegel-Zawadzka, Malgorzata
Szoltysek, Katarzyna
Valentová, Helena
Vella, Anthony

Cosmetics

Budde, Jürgen
Dos Santos Baptista, Bráulio
Eisenbrand, Gerhard

Ekar, Igor
Ferrara, Lydia
Filip, Vladimir

Golja, Viviana
Henle, Thomas
Ivanov, Kalintcho

Jirovetz, Leopold
Koskenkorva, Anneli
Kroyer, Gerhard Th.
Lalljie, Sam
Panovská, Zdenka
Pearce, Steven
Plocková, Milada

Rihakova, Zdenka
Rollin, Patrick
Schäfer, Karola
Schwack, Wolfgang
Sime, John
Studer, Alfred
Szoltysek, Katarzyna

Topalova, Ivanka Chrstova
Uccella, Nicola Antonio
Valentová, Helena
Whitehead, John A.
Williams, Peter
Zoric, Andreja

Dairy

Aalbersberg, Willem Y.
Ager, Elaine
Alder, Lutz
Anderssen, Valborg
Andrews, Anthony
Anklam, Elke
Applebye, Ulla
Barron, Luis Javier Rodríguez
Bazulic, Davorin
Birlouez, Inès
Bosset, Jacques Olivier
Brathen, Gudmund
Brechany, Elizabeth
Brunn, Hubertus
Burini, Giovanni
Calvo, Marta Maria
Calzolari, Claudio
Cerkvenik, Vesna
Chroneos, Ioannis
Chrzanowska, Jozefa
Chumchalová, Jana
Clark, David
Coveney, Leslie
Curda, Ladislav
Dawihl, Gerd
De Block, Jan
De Jong, Jacob
De Kruif, Kees C.G.
De Ruiter, Gerhard A.
De Ruyck, Hendrik
Decaris, Bernard
Di Luccia, Aldo
Dickinson, Eric
Dirinck, Patrick
Doganoc, Darinka Zdenka
Dolezal, Marek
Dossena, Arnaldo
Dziuba, Jerzy
Dzwolak, Waldemar
Ekar, Igor
Ellen, Geert
Estelecki, Ilona
Evangelisti, Filippo
Favretto, Luciano
Filip, Vladimir
Fillery-Travis, Annette

Finney, Graham
Frias, Juana
Gabrielli Favretto, Luciana
Gaucheron, Frédéric
Guillen, M. D.
Hampton, Ian
Hardi, Jovica
Heinonen, Marina
Henle, Thomas
Herman, Lieve
Holmer, Gunhild
Horne, David S.
Ilsbroux, Ingrid
Imbs, Boguslaw
Jägerstad, Margaretha
Jones, Arthur David
Jones, James
Juarez, Manuela
Keurulainen, Ritva
Klostermeyer, Henning
Korhonen, Hann
Kostyra, Henryk
Krause, Wolfgang
Krska, Rudolf
Lampolahti, Soili
Leibetseder, Josef
Leigh, Anthony
Luf, Wolfgang
Maclean, Wim
Malwitz, Dietmar
Marini, Domenico
Martelli, Aldo
Martinez-Castro, Isabel
Matilainen, Katri
Mills, Elizabeth Naomi Clare
Munck, Lars
Murray, Brent Stuart
Nöhle, Ulrich
Northolt, Martin
Novakovic, Predrag
Nursten, Harry Erwin
O'Brien, John
Olano, Agustin
Olieman, Cornelis
Otte, Jeanette
Pavelka, Jiri

Pertoldi Marletta, Giuliana
Petz, Michael
Pischetsrieder, Monika
Pisciotta, Gennaro
Plocková, Milada
Rahali, Véronique
Rantamäkki, Pirjio
Restani, Patrizia
Rimkus, Gerhard G.
Salminen, Seppo Jaakko
Sapunar-Postruznik, Jasenka
Sarpeid, Hans-Jacob
Schmidt, Heinz
Schmolck, Werner
Schreyen, Luc
Schutte, Leonard
Sforza, Stefano
Shackleton, Ronald
Sherlock, John C.
Siezen, Roland J.
Sinigoj-Gacnik, Ksenija
Skibsted, Leif H.
Sorhaug, Terje
Stark, Jacques
Stetina, Jiri
Suiraková, Eva
Thomas, Mark Andrew
Llewellyn
Timmermans, Eric
Toikkanen, Kari
Tömösközi, Sándor
Tykkyläinen, Paavo
Ulberth, Franz
Van Boekel, Tiny
Van Dael, Peter
Van Eckert, Renate
Van Renterghem, Roland
Van Rhyn, Hans
Wiedner, Peter
Wilde, Peter
Wilson, Philip
Wojtatowicz, Maria
Zagt, Robert
Ziajka, Stefan
Ziino, Marisa
Zunin, Paola

Drugs

Bazulic, Davorin
Brunn, Hubertus
Büning-Pfaue, Hans
Cerkvenik, Vesna
De Brabander, Hubert
De Jong, Jacob
De Ruyck, Hendrik
Ibe, Frank
Jirovetz, Leopold

Kan, Kees (C.A.)
Kaufmann, Anton
Koch, Herbert
Lalljie, Sam
Lehtonen, Pekka
Malisch, Rainer
Österdahl, Bengt-Göran
Petz, Michael
Ruiter, Adriaan

Schlatter, Christian
Schmidt, Heinz
Schrenk, Dieter
Sinigoj-Gacnik, Ksenija
Stark, Jacques
Ternes, Waldemar
Valentini, Giuseppa
Van Rhyn, Hans
Van Vyncht, Gery

Drying

Cooper, Julian
Di Luccia, Aldo
Dos Santos Baptista, Bráulio
Hegedusic, Vesna
Henle, Thomas
Horne-Ekman, Maarit
Iciek, Jan
Ilari, Jean-Luc

Keurulainen, Ritva
Kmiecik, Waldemar Andrzej
Lisiewska, Zofia Barbara
Lisinska, Grazyna
Löliger, J.
Marin, Michèle
Pezacki, Wincenty
Pipek, Petr

Roos, Yrjö Henrik
Rutledge, Douglas
Sereno, Alberto M.C.
Steinbuch, Erwin
Studer, Alfred
Timmermans, Eric
Von Wietersheim, Eugen

Eggs

Alder, Lutz
Bazulic, Davorin
Brockmann, Rainer
Brunn, Hubertus
Büning-Pfaue, Hans
Cerkvenik, Vesna
De Ruyck, Hendrik
Doganoc, Darinka Zdenka
Dolezal, Marek
Groothuis, Dirk G.

Herman, Lieve
Huopalahti, Rainer
Ingr, Ivo
Järvenpää, Eila
Kan, Kees (C.A.)
Leibetseder, Josef
Littmann-Nienstedt, Sigrid
Luf, Wolfgang
Mikova, Kamila
Petz, Michael

Rimkus, Gerhard G.
Sapunar-Postruznik, Jasenka
Schmidt, Heinz
Sinigoj-Gacnik, Ksenija
Smolinska, Teresa
Ternes, Waldemar
Trojan, Marek
Uijttenboogaart, Theo
Van Den Bosch, Gerrit
Van Rhyn, Hans

Electrochemistry

Amaro Pinto, Rui Manuel
Bontempelli, Gino
Botrè, Francesco
Campanella, Luigi
Enders, Peter W.

Hasenay, Damir
Hils, Arno K. A.
Luf, Wolfgang
Mazzei, Franco
Popov, Dimitre

Russell, Wendy Roslyn
Sontag, Gerhard
Varadi, Maria
Vidal-Valverde, Concepcion

Electrophoresis

Amaro Pinto, Rui Manuel
Andrews, Anthony
Bachman, Stefania
Basman, Arzu
Bauer, Friedrich
Beljaars, Paul
Benda, Vladimir

Berardo, Nicola
Betsche, Thomas
Bhat, Mahalingeshwara
Calvo, Marta Maria
Celik, Süeda
Chroneos, Ioannis
Chumchalová, Jana

Clark, David
Collar Esteve, Concha
De Block, Jan
Decaris, Bernard
Di Luccia, Aldo
Dziuba, Jerzy
Dzwolak, Waldemar

Enders, Peter W.
Evidente, Antonio
Glück, Bernfried
Goodall, David M.
Gramshaw, J.W.
Hauser, Eugen J. B.
Henle, Thomas
Heredia, Antonia
Jones, Arthur David
Knieling, Ralph G.
Köksel, Hamit
Kostyra, Henryk
Kovacs, Elisabeth Teresia
Kozlowska, Halina
Krause, Wolfgang
Krkoskova, Bernadetta
Kroll, Jürgen

Kroon, Paul
Kucera, Jiri
Kvasnicka, Frantisek
Lalljie, Sam
Luneia, Roberto
Melzoch, Karel
Mills, Elizabeth Naomi Clare
Murray, Brent Stuart
Pezacki, Wincenty
Plocková, Milada
Rahali, Véronique
Restani, Patrizia
Robak, Malgorzata
Rodziewicz, Anna
Rohn, Sascha
Sarpeid, Hans-Jacob
Sass-Kiss, Agnes

Schwenke, Klaus Dieter
Sorensen, Susanne
Stetina, Jiri
Suiraková, Eva
Szoltysek, Katarzyna
Tauscher, Bernhard
Trojan, Marek
Troy, Declan J.
Ugarcic-Hardi, Zaneta
Van Eckert, Renate
Vieths, Stefan
Weder, Jürgen, K.P.
Wijngaards, Gerrit
Witkowska, Danuta
Yalcin, Erkan
Zyla, Krzysztof

Enzymology

Adams, J. Brian
Andrews, Anthony
Basman, Arzu
Berger, Ralf Günter
Betsche, Thomas
Bhat, Mahalingeshwara
Chrzanowska, Jozefa
Clark, David
Dalev, Pencho
Damiani, Pietro
De Ruiter, Gerhard A.
Demopolous, Constantinos A.
Di Luccia, Aldo
Dziuba, Jerzy
Eerikäinen, Tero
Eisenbrand, Gerhard
Faulds, Craig
Iori, Renato
Kas, Jan

Klostermeyer, Henning
Köksel, Hamit
Krause, Wolfgang
Kroon, Paul
Kucera, Jiri
Long, Alan
Melchior Larsen, Lone
Meunier, Jean-Claude
Moreira Da Silva, Aida Maria
Otte, Jeanette
Pearce, J.
Pearce, Steven
Rauch, Pavel
Robak, Malgorzata
Rodziewicz, Anna
Rogerson, Frank
Salgò, Andràs
Sarpeid, Hans-Jacob
Siezen, Roland J.

Sime, John
Slominska, Lucyna
Sorensen, Hilmer
Sorensen, Susanne
Sorhaug, Terje
Szoltysek, Katarzyna
Timmermans, Eric
Troy, Declan J.
Vodrazka, Zdenek
Voragen, Fons
Weder, Jürgen, K.P.
Wijngaards, Gerrit
Williamson, Gary
Witkowska, Danuta
Wucherpfennig, Karl
Yalcin, Erkan
Zabetakis, Ioannis
Zyla, Krzysztof

Ethanol

Czarnecki, Zbgniew
Dietrich, Helmut
Enders, Peter W.
Knieling, Ralph G.
Lehtonen, Pekka
Lesniak, Wladyslaw

Liddle, Peter
Marin, Michèle
Marini, Domenico
Melzoch, Karel
Mosandl, Armin
Nowak, Jacek

Rizov, Nicolay
Rychtera, Mojmír
Stevens, Roger
Topalova, Ivanka Chrstova
Vischer, Michaela
Webb, Colin

Fat replacers

Corradini, Claudio
De Ruiter, Gerhard A.
Fillery-Travis, Annette
Gertz, Christian
Gunstone, Frank

Ilari, Jean-Luc
Krkoskova, Bernadetta
Matissek, Reinhard
Novakovic, Predrag
Otte, Jeanette

Pezacki, Wincenty
Pollmer, Udo
Ragotzky, Klaus
Rops, Wichard
Schwarz, Walter

Steegmans, Monique
Studer, Alfred

Suhaj, Milan
Troy, Declan J.

Tsanev, Roumen
Tykkyläinen, Paavo

Fermentation

Aalbersberg, Willem Y.
Accorsi, Carla Alberta
Balduck, Paul
Barron, Luis Javier Rodríguez
Benedito, Carmen
Berger, Ralf Günter
Betsche, Thomas
Böhm, Josef
Bosset, Jacques Olivier
Calvo, Marta Maria
Carder, John
Christoph, Norbert
Chumchalová, Jana
Clark, David
Collar Esteve, Concha
Czarnecka, Maria
Czarnecki, Zbgniew
Eerikäinen, Tero
Evidente, Antonio
Frias, Juana
Fryer, John
Gonzalez-Sanjose, Maria Luisa
Hamilton, Colin A.

Hardi, Jovica
Hauser, Eugen J. B.
Henle, Thomas
Home, Silja
Horne-Ekman, Maarit
Ilsbroux, Ingrid
Jirovetz, Leopold
Khokhar, Santosh
Klupsch, Robert N.
Krause, Wolfgang
Lesniak, Wladyslaw
Löliger, J.
Martinez Anaya, Antonia M.
Melzoch, Karel
Miskiewicz, Tadeusz
Northolt, Martin
Novakovic, Predrag
Nowak, Jacek
Pietkiewicz, Jerzy Jan
Pipek, Petr
Pisciotta, Gennaro
Plocková, Milada
Räsänen, Janne

Rihar Jereb, Bernarda
Roel, Peter
Rogerson, Frank
Rollin, Patrick
Rombouts, Frank
Rychtera, Mojmír
Rymowicz, Waldemar
Sandberg, Ann-Sofie
Schmitt, Rudolf
Schreyen, Luc
Sorhaug, Terje
Steinbuch, Erwin
Stempiewicz, Regina
Stevens, Roger
Suiraková, Eva
Ugarcic-Hardi, Zaneta
Van Den Bosch, Gerrit
Vidal-Valverde, Concepcion
Webb, Colin
Wilson, Peter D.G.
Wojtatowicz, Maria
Wood, Brian J.B.
Zyla, Krzysztof

Dietary fibre

Aura, Anna-Marija
Basman, Arzu
Beljaars, Paul
Betsche, Thomas
Bhat, Mahalingeshwara
Bjergegaard, Charlotte
Böhm, Josef
Branch, Simon
Coimbra, Manuel António
Cooper, Julian
Corradini, Claudio
Dalev, Pencho
Deweghe, Liane
Estelecki, Ilona
Frias, Juana
Heredia, Antonia
Hodgson, Ian

Horváth-Mosonyi, Magda
Howard, Julie
Kmiecik, Waldemar Andrzej
Köksel, Hamit
Krkoskova, Bernadetta
Lindhauer, Meinolf G.
Lisiewska, Zofia Barbara
Lisinska, Grazyna
Mandic, Milena L.
Martelli, Aldo
Mischnick, Petra
Niemi, Sanna-Maria
Novakovic, Predrag
O' Neill, Ian
Popken, Anne M.
Pospisil, Jasna
Ragotzky, Klaus

Räsänen, Janne
Saarinen, Niina
Sandberg, Ann-Sofie
Seibel, Wilfried
Sorensen, Hilmer
Steegmans, Monique
Steinhart, Hans
Szilágyi, Szilárd
Tömösközi, Sándor
Tykkyläinen, Paavo
Ugarcic-Hardi, Zaneta
Van Dokkum, Wim
Vidal-Valverde, Concepcion
Vokk, Raivo
Waldron, Keith
Zyla, Krzysztof

Filtration

Bubnik, Zdenek
Calvo, Marta Maria
Cooper, Julian
Dietrich, Helmut
Mac Dowall, James

Marin, Michèle
Melzoch, Karel
Pozderovic, Andrija
Rihar Jereb, Bernarda
Shackleton, Ronald

Timmermans, Eric
Wang, Rui
Wucherpfennig, Karl

Fish

Aalbersberg, Willem Y.
Alder, Lutz
Aro, Tarja
Aubourg, Santiago
Balling Engelsen, Soren
Bazulic, Davorin
Beddows, Clifford G.
Botrè, Francesco
Brathen, Gudmund
Brockmann, Rainer
Brunn, Hubertus
Bykowski, Piotr Jan
Cerkvenik, Vesna
Christophersen, Carsten
Clark, Michael
Damiani, Pietro
Demopolous, Constantinos A.
Di Natale, Corrado
Doganoc, Darinka Zdenka
Empis, José
Enders, Peter W.
Farmer, Linda
Favretto, Luciano
Finney, Graham
Gabrielli Favretto, Luciana
Gil, Ana Maria Pissarra C.
Glück, Bernfried
Gojmerac, Tihomira
Groothuis, Dirk G.
Guillen, M. D.
Holmer, Gunhild
Ilmoja, Kalle
Ingr, Ivo
Jones, Arthur David
Koch, Herbert
Kucera, Jiri
Lalljie, Sam
Lampolahti, Soili
Lapveteläinen, Anja
Leibetseder, Josef
Luckas, Bernd
Malwitz, Dietmar
Munck, Lars
Nehring, Ulrich P.
Nortvedt, Ragnar
Pastoriza, Laura
Pavelka, Jiri
Pertoldi Marletta, Giuliana
Pipek, Petr
Rimkus, Gerhard G.
Ruiter, Adriaan
Salvo, Francesco
Sapunar-Postruznik, Jasenka
Sarpeid, Hans-Jacob
Schmidt, Heinz
Sikorski, Zdzistaw
Sinigoj-Gacnik, Ksenija
Surowka, Krzysztof
Thornton, Raymond E.
Van Rhyn, Hans
Weisz, Richard
Zabetakis, Ioannis
Ziino, Marisa

Food composition

Andrews, Anthony
Anklam, Elke
Assimakopoulou, Angelique
Aubourg, Santiago
Aura, Anna-Marija
Baines, David Allan
Balling Engelsen, Soren
Baltes, Werner
Barron, Luis Javier Rodríguez
Basman, Arzu
Battaglia, Reto
Beernaert, Hedwig
Beljaars, Paul
Benedito, Carmen
Betsche, Thomas
Bjergegaard, Charlotte
Boskou, Dimitrius
Bosset, Jacques Olivier
Branch, Simon
Brechany, Elizabeth
Brown, Peter Anthony
Bubnik, Zdenek
Calvo, Marta Maria
Campbell, Duncan J.
Cejpek, Karel
Celik, Süeda
Coimbra, Manuel António
Corradini, Claudio
Corvi, Claude Albert
Coveney, Leslie
Dalev, Pencho
Damiani, Pietro
Davies, Alan Philipp
Davies, Robert John
De Ruyck, Hendrik
Deelstra, Hendrik
Deweghe, Liane
Di Luccia, Aldo
Dietrich, Helmut
Doganoc, Darinka Zdenka
Dolezal, Marek
Drdak, Milan
Eberle, Mike
Ennion, Ronald
Evangelisti, Filippo
Evidente, Antonio
Favretto, Luciano
Fenwick, Gruffydd Roger
Ferrara, Lydia
Filip, Vladimir
Frias, Juana
Fryer, John
Gabrielli Favretto, Luciana
Garncarek, Barbara
Gatti, Gian Carlo
Gaucheron, Frédéric
Gertz, Christian
Gladovic, Natasa
Glück, Bernfried
Golachowski, Antoni
Gonzalez-Sanjose, Maria Luisa
Gordon, Michael
Gramshaw, J.W.
Grenby, Trevor Hilary
Guillen, M. D.
Hägg, Margareta
Häkkinen, Sari
Hardi, Jovica
Hauser, Eugen J. B.
Henle, Thomas
Heredia, Antonia
Hermus, Rudolph J.J.
Herraiz Tomico, Tomas
Hietaniemi, Veli
Hitchcock, Christopher
Holandová, Katerina
Holasova, Marie
Honikel, Karl Otto
Hood, Ted D.E.
Horne, David S.
Horváth-Mosonyi, Magda
Horvatic, Marija
Hrncirik, Karel
Hruskar, Mirjana
Huopalahti, Rainer
Ilmoja, Kalle
Ivanov, Kalintcho
Jarmoluk, Andrzej
Järvenpää, Eila
Jirovetz, Leopold
Jörissen, Urban
Juarez, Manuela
Kalac, Pavel
Kallio, Heikki
Kedzior, Wladyslaw
Kerkvliet, Jacob
Khokhar, Santosh
Kivistö, Laura
Klupsch, Robert N.

Kmiecik, Waldemar Andrzej
Köksel, Hamit
Komaitis, Michael
Kovac, Milan
Kovacs, Elisabeth Teresia
Kovatcheva-Apostolova, Elena
Kozlowska, Halina
Krala, Lucjan
Kroll, Jürgen
Kucera, Jiri
Lasztity, Radomir
Lees, Ronald
Lehtonen, Pekka
Leibetseder, Josef
Leigh, Anthony
Lesniak, Wladyslaw
Leszczynski, Waclaw
Lisiewska, Zofia Barbara
Lisinska, Grazyna
Long, Alan
Luf, Wolfgang
Määttä, Kaisu
Makris, Dimitris
Malwitz, Dietmar
Mandic, Milena L.
Marini, Domenico
Marioleas, Panagiotis
Martinez-Castro, Isabel
Masková, Eva
Matilainen, Katri
Matissek, Reinhard
Matuszek, Tadeusz
Melzoch, Karel
Mikova, Kamila
Molnar-Perl, Ibolya
Moreira Da Silva, Aida Maria
Mörsel, Jörg-Thomas
Mottram, Donald
Murkovic, Michael
Nortvedt, Ragnar
Noteborn, Hubert
Novakovic, Predrag
Olano, Agustin
Olieman, Cornelis
Ollilainen, Velimatti
Ooghe, Wilfried
Otte, Jeanette

Palasinski, Mieczyslaw
Panovská, Zdenka
Parr, Adrian James
Pastoriza, Laura
Pavelka, Jiri
Payne, Nigel Kenneth
Pearce, J.
Pearce, Steven
Peksa, Anna
Pertoldi Marletta, Giuliana
Pezacki, Wincenty
Pfalzgraf, Andreas
Pietkiewicz, Jerzy Jan
Piironen, Vieno
Pilizota, Vlasta
Pisciotta, Gennaro
Pokorny, Jan
Preuss, Axel
Primorac, Ljiljana
Prugar, Jaroslav
Pudil, Frantisek
Rahali, Véronique
Ribarova, Fanny
Rihakova, Zdenka
Ristow, Reinhard
Rizov, Nicolay
Roedig-Penman, Andrea
Rogerson, Frank
Roos, Yrjö Henrik
Rops, Wichard
Ruiter, Adriaan
Rutkowski, Antoni
Sakellariou, Christina
Salgò, Andràs
Salvo, Francesco
Saukko, Maire
Schulte, Erhard
Schwack, Wolfgang
Schwenke, Klaus Dieter
Sforza, Stefano
Simon-Sarkadi, Livia
Sinigoj-Gacnik, Ksenija
Skibniewska, Krystyna A.
Skrabka-Blotnicka, Teresa
Skrökki, Leila Anneli
Smolinska, Teresa
Sorensen, Hilmer

Sorensen, Susanne
Sousa, Isabel
Spiro, Michael
Steegmans, Monique
Stevens, Roger
Studer, Alfred
Suhaj, Milan
Surowka, Krzysztof
Suutarinen, Marjaana
Szilágyi, Szilárd
Talbot, Geoff
Tateo, Fernando
Ternes, Waldemar
Thier, Hans-Peter
Thompson, Kenneth Clive
Toivo, Jari
Tömösközi, Sándor
Tóth-Markus, Marianna
Tsanev, Roumen
Uccella, Nicola Antonio
Ugarcic-Hardi, Zaneta
Uijttenboogaart, Theo
Ukmar Maljevac, Damjana
Vahcic, Nada
Vahteristo, Liisa
Valentová, Helena
Van Den Bosch, Gerrit
Van Renterghem, Roland
Van Vyncht, Gery
Varadi, Maria
Velisek, Jan
Venskutonis, Petras Rimantas
Vidal-Valverde, Concepcion
Vischer, Michaela
Weder, Jürgen, K.P.
Wijngaards, Gerrit
Wilska-Jeszka, Jadwiga
Wilson, Philip
Wilson, Reginald
Yanishlieva-Maslarova, Nedjalka
Zegota, Henryk
Zidaric, Metka
Ziino, Marisa
Zunin, Paola

Fouling and cleaning

De Kruif, Kees C.G.
Fisher, Leonard
Matuszek, Tadeusz

Melzoch, Karel
Pezacki, Wincenty
Plocková, Milada

Rodziewicz, Anna
Storgards, Erna
Studer, Alfred

Freezing

Aubourg, Santiago
Bykowski, Piotr Jan

Hajduk, Ewa
Häkkinen, Sari

Hegedusic, Vesna
Hood, Ted D.E.

Khokhar, Santosh
Kmiecik, Waldemar Andrzej
Krala, Lucjan
Kramer, Jörg
Lisiewska, Zofia Barbara
Marin, Michèle
Pezacki, Wincenty

Pozderovic, Andrija
Räsänen, Janne
Roos, Yrjö Henrik
Sereno, Alberto M.C.
Sikorski, Zdzistaw
Steinbuch, Erwin
Studer, Alfred

Suiraková, Eva
Surowka, Krzysztof
Suutarinen, Marjaana
Tomasik, Piotr
Uijttenboogaart, Theo

Fruits and vegetables

Adams, J. Brian
Alder, Lutz
Ashurst, Philip Roy
Aura, Anna-Marija
Beddows, Clifford G.
Berger, Ralf Günter
Birlouez, Inès
Böhm, Volker
Bontempelli, Gino
Brown, Peter Anthony
Brunn, Hubertus
Budde, Jürgen
Buglass, Alan J.
Büning-Pfaue, Hans
Catalá, Ramon
Cejpek, Karel
Christophersen, Carsten
Clark, Michael
Coimbra, Manuel António
Corradini, Claudio
Delgadillo, Ivonne
Dettweiler, Gerd
Di Natale, Corrado
Dietrich, Helmut
Dirinck, Patrick
Drdak, Milan
Duran, Luis
Eberle, Mike
Eshuis, Dolf F.
Farkas, Joszef
Faulds, Craig
Favretto, Luciano
Fenwick, Gruffydd Roger
Ferrara, Lydia
Finney, Graham
Frias, Juana
Galluser, Anita
Garncarek, Barbara
Gil, Ana Maria Pissarra C.
Gonzalez-Sanjose, Maria Luisa
Goodman, Bernard
Hägg, Margareta
Hakala, Mari
Häkkinen, Sari
Hanewinkel-Meshkini, Susanne
Harris, Caroline A.
Hegedusic, Vesna
Heinonen, Marina
Heinzler, Matthias

Henshall, David
Heredia, Antonia
Hills, Brian
Hils, Arno K. A.
Holandová, Katerina
Holst, Birgit
Hopia, Anu
Horne-Ekman, Maarit
Horváth-Mosonyi, Magda
Hrncirik, Karel
Huopalahti, Rainer
Ilmoja, Kalle
Iori, Renato
Jägerstad, Margaretha
Järvenpää, Eila
Jirovetz, Leopold
Kallio, Heikki
Kerkvliet, Jacob
Kmiecik, Waldemar Andrzej
Koskenkorva, Anneli
Kozlowska, Halina
Kramer, Jörg
Kroon, Paul
Kvasnicka, Frantisek
Lach, Günter
Lapveteläinen, Anja
Lisiewska, Zofia Barbara
Lisinska, Grazyna
Lugasi, Andrea
Määttä, Kaisu
Malbon, Raymond
Marchelli, Rosangela
Marioleas, Panagiotis
Martin, Gérard
Marx, Friedhelm
Melchior Larsen, Lone
Molnar, Pal
Molnar-Perl, Ibolya
Mosandl, Armin
Mottram, Donald
Mraz, Igor
Nehring, Ulrich P.
Neicheva, Anastasia
Noteborn, Hubert
O' Neill, Ian
Ochs, Stefan
Olano, Agustin
Ooghe, Wilfried
Österdahl, Bengt-Göran
Parr, Adrian James

Pearce, J.
Pfannhauser, Werner
Pilizota, Vlasta
Pisciotta, Gennaro
Popken, Anne M.
Pospisil, Jasna
Poustka, Jan
Pozderovic, Andrija
Prugar, Jaroslav
Reglero, Guillermo J.
Rihar Jereb, Bernarda
Roel, Peter
Rollin, Patrick
Roos, Yrjö Henrik
Russell, Wendy Roslyn
Saarinen, Niina
Sakellariou, Christina
Sass-Kiss, Agnes
Saukko, Maire
Schreier, Peter
Schulte, Erhard
Schwack, Wolfgang
Sereno, Alberto M.C.
Sherlock, John C.
Siebers, Johannes
Siegmund, Barbara
Smith, Linda Bernhardine Margaret
Steinbuch, Erwin
Steiner, Ingrid
Subaric, Drago
Suutarinen, Marjaana
Szilágyi, Szilárd
Talou, Thierry
Tauscher, Bernhard
Thier, Hans-Peter
Thornton, Raymond E.
Toikkanen, Kari
Tóth-Markus, Marianna
Van Boekel, Tiny
Van Der Schee, Henk A.
Venskutonis, Petras Rimantas
Vieths, Stefan
Voragen, Fons
Waldron, Keith
Weisz, Richard
Wiggins, Edgar Hugh
Williamson, Gary
Wilska-Jeszka, Jadwiga
Wilson, Reginald

Winkeler, Heinz-Dieter
Winterhalter, Peter

Wucherpfennig, Karl
Zabetakis, Ioannis

Zegota, Henryk
Zidaric, Metka

Gas chromatography

Adamantiadou, Sophia
Ager, Elaine
Alder, Lutz
Ames, Jennifer
Anklam, Elke
Baigrie, Brian
Baltes, Werner
Barron, Luis Javier Rodríguez
Bauer, Ulrich
Bazulic, Davorin
Beernaert, Hedwig
Beljaars, Paul
Berardo, Nicola
Berger, Ralf Günter
Betsche, Thomas
Böhm, Josef
Bosset, Jacques Olivier
Brathen, Gudmund
Brechany, Elizabeth
Brockmann, Rainer
Brown, Peter Anthony
Buglass, Alan J.
Büning-Pfaue, Hans
Burini, Giovanni
Bykowski, Piotr Jan
Calvo, Marta Maria
Cejpek, Karel
Cerkvenik, Vesna
Christoph, Norbert
Chroneos, Ioannis
Clark, David
Clutton, David
Coimbra, Manuel António
Collar Esteve, Concha
Copíková, Jana
Corvi, Claude Albert
Cotte, Virginie
Crossy, Paul
Culik, Jirí
Damiani, Pietro
Davidek, Jiri
De Brabander, Hubert
Dettweiler, Gerd
Deweghe, Liane
Di Luccia, Aldo
Dietrich, Helmut
Dirinck, Patrick
Dolezal, Marek
Dos Santos Baptista, Bráulio
Ducauze, Christian J.
Ehlers, Dorothea
Eisenbrand, Gerhard
Ellen, Geert
Enders, Peter W.
Evangelisti, Filippo

Evidente, Antonio
Farkas, Pavel
Farmer, Linda
Filip, Vladimir
Finnegan, Derek
Gabrielli Favretto, Luciana
Gates, Leonard Michael
Gertz, Christian
Gilbert, John
Glück, Bernfried
Gordon, Michael
Graille, Jean
Gramshaw, J.W.
Guillen, M. D.
Gunstone, Frank
Hahn, Harald
Hakala, Mari
Häkkinen, Sari
Hardi, Jovica
Hauser, Eugen J. B.
Heinzler, Matthias
Henle, Thomas
Heredia, Antonia
Herraiz Tomico, Tomas
Hietaniemi, Veli
Holandová, Katerina
Holmer, Gunhild
Howard, Julie
Hrncirik, Karel
Huopalahti, Rainer
Ibe, Frank
Ilmoja, Kalle
Jägerstad, Margaretha
Järvenpää, Eila
Jirovetz, Leopold
Jones, Alan
Jones, James
Jörissen, Urban
Juarez, Manuela
Kalac, Pavel
Kallio, Heikki
Klein, Bernard
Klein, Erich
Knieling, Ralph G.
Koch, Herbert
Komaitis, Michael
Kontominas, Michael
Kovatcheva-Apostolova, Elena
Kozlowska, Halina
Kroh, Lothar W.
Krska, Rudolf
Kuusisto, Päivi
Laakso, Päivi
Lach, Günter
Lalljie, Sam

Lambert, Michael
Lampi, Anna-Maija
Lehtonen, Pekka
Leitner, Erich
Liddle, Peter
Luckas, Bernd
Luf, Wolfgang
Luneia, Roberto
Määttä, Kaisu
Marioleas, Panagiotis
Martinez-Castro, Isabel
Marx, Friedhelm
Matissek, Reinhard
Mc Donald, Mark
Melzoch, Karel
Mikova, Kamila
Mischnick, Petra
Mlotkiewicz, Jerzy
Molnar-Perl, Ibolya
Moret, Ivo
Mörsel, Jörg-Thomas
Mosandl, Armin
Mottram, Donald
Neicheva, Anastasia
Nursten, Harry Erwin
Obretonov, Tzvetan
Ochs, Stefan
Ogaard Madsen, Jorgen
Olano, Agustin
Österdahl, Bengt-Göran
Parker, Jane
Pertoldi Marletta, Giuliana
Petz, Michael
Pfannhauser, Werner
Pilizota, Vlasta
Pischetsrieder, Monika
Popken, Anne M.
Poustka, Jan
Pozderovic, Andrija
Primorac, Ljiljana
Pudil, Frantisek
Restani, Patrizia
Ridgway, Christopher
Rimkus, Gerhard G.
Ristow, Reinhard
Rizov, Nicolay
Roedig-Penman, Andrea
Rogerson, Frank
Roozen, Jacques
Salvo, Francesco
Schlett, Claus
Schmid, Erich R.
Schmidt, Stefan
Schreier, Peter
Schreyen, Luc

Schulte, Erhard
Scordaki, Alexandra
Sforza, Stefano
Sheppard, Peter D.
Siebers, Johannes
Siegmund, Barbara
Sikorski, Zdzistaw
Simko, Peter
Smith, Linda Bernhardine Margaret
Speer, Karl
Steegmans, Monique
Steiner, Ingrid
Subaric, Drago
Talou, Thierry

Tateo, Fernando
Tauscher, Bernhard
Ternes, Waldemar
Thornton, Raymond E.
Toivo, Jari
Topalova, Ivanka Chrstova
Tóth-Markus, Marianna
Tsanev, Roumen
Ulberth, Franz
Valentini, Giuseppa
Van Den Bosch, Gerrit
Van Der Schee, Henk A.
Van Peteghem, Carlos
Van Renterghem, Roland
Van Vyncht, Gery

Vanluchene, Eric
Velisek, Jan
Venskutonis, Petras Rimantas
Vischer, Michaela
Wang, Rui
Weisz, Richard
Williams, John Graham
Williams, Mervyn
Winkeler, Heinz-Dieter
Yanishlieva-Maslarova, Nedjalka
Zabetakis, Ioannis
Ziino, Marisa
Zoller, Otmar
Zunin, Paola

Genetic engineering

Anklam, Elke
Balduck, Paul
Beljaars, Paul
Benda, Vladimir
Berardo, Nicola
Chumchalová, Jana

Decaris, Bernard
Gunstone, Frank
Klostermeyer, Henning
Noteborn, Hubert
Nowak, Jacek
Pezacki, Wincenty

Siezen, Roland J.
Szoltysek, Katarzyna
Thornton, Raymond E.
Vieths, Stefan
Vodrazka, Zdenek
Voragen, Fons

Heterocyclic aromatic amines

Baltes, Werner
Beljaars, Paul
Herraiz Tomico, Tomas
Jägerstad, Margaretha
Jirovetz, Leopold
Jones, Arthur David

Murkovic, Michael
O' Neill, Ian
O´Brien, John
Pezacki, Wincenty
Pfannhauser, Werner
Schlatter, Christian

Sontag, Gerhard
Stanoeva, Elena
Winkeler, Heinz-Dieter
Zoller, Otmar

High pressure technology

Aura, Anna-Marija
Cheftel, Jean-Claude
Chiavaro, Emma
De Kruif, Kees C.G.
Dickinson, Eric
Dos Santos Baptista, Bráulio
Eberle, Mike
Farkas, Joszef
Gaucheron, Frédéric

Gojmerac, Tihomira
Henle, Thomas
Hietaniemi, Veli
Järvenpää, Eila
Khokhar, Santosh
Masková, Eva
Mills, Elizabeth Naomi Clare
Olano, Agustin
Pearce, Steven

Pezacki, Wincenty
Skibsted, Leif H.
Tauscher, Bernhard
Ternes, Waldemar
Tomasik, Piotr
Troy, Declan J.
Zabetakis, Ioannis

Hormones

Beernaert, Hedwig
Beljaars, Paul
Cerkvenik, Vesna
De Brabander, Hubert
De Jong, Jacob
Eisenbrand, Gerhard

Gojmerac, Tihomira
Huopalahti, Rainer
Ibe, Frank
Koch, Herbert
Lalljie, Sam
Malisch, Rainer

Mc Donald, Mark
Morgan, Michael R.A.
Noteborn, Hubert
Saarinen, Niina
Schlatter, Christian
Schlett, Claus

Schmid, Erich R.
Schrenk, Dieter
Skibniewska, Krystyna A.

Sontag, Gerhard
Van Peteghem, Carlos
Van Rhyn, Hans

Van Vyncht, Gery

HPLC

Accorsi, Carla Alberta
Adamantiadou, Sophia
Ager, Elaine
Alder, Lutz
Amarowicz, Ryszard
Ames, Jennifer
Andrews, Anthony
Anklam, Elke
Aust, Olivier
Baigrie, Brian
Barron, Luis Javier Rodríguez
Bauer, Friedrich
Bazulic, Davorin
Beddows, Clifford G.
Beernaert, Hedwig
Beljaars, Paul
Benedito, Carmen
Berardo, Nicola
Berger, Ralf Günter
Betsche, Thomas
Bhat, Mahalingeshwara
Birlouez, Inès
Böhm, Josef
Brathen, Gudmund
Brechany, Elizabeth
Brockmann, Rainer
Brown, Peter Anthony
Bubnik, Zdenek
Budde, Jürgen
Buglass, Alan J.
Büning-Pfaue, Hans
Burini, Giovanni
Bykowski, Piotr Jan
Calvo, Marta Maria
Calzolari, Claudio
Campbell, Duncan J.
Cejpek, Karel
Celik, Süeda
Cerkvenik, Vesna
Chiavaro, Emma
Chroneos, Ioannis
Chrysafidis, Dimitrios
Chumchalová, Jana
Clark, David
Collar Esteve, Concha
Cooper, Julian
Copíková, Jana
Corradini, Claudio
Corvi, Claude Albert
Culik, Jirí
Dalev, Pencho
Damiani, Pietro
Davidek, Jiri
De Block, Jan

De Brabander, Hubert
De Jong, Jacob
De Ruiter, Gerhard A.
De Ruyck, Hendrik
Dechow, Arndt
Demopolous, Constantinos A.
Dettweiler, Gerd
Deweghe, Liane
Di Luccia, Aldo
Dietrich, Helmut
Dirinck, Patrick
Doncheva, Ivanka
Dossena, Arnaldo
Dziuba, Jerzy
Ehlers, Dorothea
Eisenbrand, Gerhard
Ekar, Igor
Enders, Peter W.
Erning, Dieter
Evangelisti, Filippo
Evidente, Antonio
Farmer, Linda
Faulds, Craig
Feinberg, Max
Ferrara, Lydia
Filip, Vladimir
Frias, Juana
Garncarek, Barbara
Gaucheron, Frédéric
Gertz, Christian
Gilbert, John
Gladovic, Natasa
Glück, Bernfried
Gojmerac, Tihomira
Goodall, David M.
Gordon, Michael
Graille, Jean
Gramshaw, J.W.
Grenby, Trevor Hilary
Grenby, Trevor Hilary
Hägg, Margareta
Häkkinen, Sari
Hanewinkel-Meshkini, Susanne
Hauser, Eugen J. B.
Heinonen, Marina
Heinzler, Matthias
Henle, Thomas
Heredia, Antonia
Herraiz Tomico, Tomas
Hietaniemi, Veli
Hils, Arno K. A.
Holandová, Katerina
Holasova, Marie

Holmer, Gunhild
Holst, Birgit
Hopia, Anu
Howard, Julie
Hrncirik, Karel
Huopalahti, Rainer
Ibe, Frank
Ilmoja, Kalle
Iori, Renato
Jägerstad, Margaretha
Jarmoluk, Andrzej
Järvenpää, Eila
Jirovetz, Leopold
Jones, Alan
Jones, Arthur David
Jones, James
Jörissen, Urban
Kallio, Heikki
Kaufmann, Anton
Khokhar, Santosh
Klein, Bernard
Klein, Erich
Kmiecik, Waldemar Andrzej
Knieling, Ralph G.
Koch, Herbert
Komaitis, Michael
Kombal, Ralph
Kontominas, Michael
Kostyra, Henryk
Kovatcheva-Apostolova, Elena
Kozlowska, Halina
Krause, Wolfgang
Kroh, Lothar W.
Kroll, Jürgen
Kroon, Paul
Krska, Rudolf
Kucera, Jiri
Kuusisto, Päivi
Kvasnicka, Frantisek
Laakso, Päivi
Lalljie, Sam
Lambert, Michael
Lampi, Anna-Maija
Lehtonen, Pekka
Lisiewska, Zofia Barbara
Luckas, Bernd
Luf, Wolfgang
Lugasi, Andrea
Luneia, Roberto
Määttä, Kaisu
Mäkinen, Marjukka
Makris, Dimitris
Marchelli, Rosangela
Marini, Domenico

Marioleas, Panagiotis
Martelli, Aldo
Martinez Anaya, Antonia M.
Matissek, Reinhard
Mc Donald, Mark
Melzoch, Karel
Meunier, Jean-Claude
Mikova, Kamila
Mikuschka, Gerhard
Mills, Elizabeth Naomi Clare
Mlotkiewicz, Jerzy
Moilanen, Raija
Molnar-Perl, Ibolya
Mörsel, Jörg-Thomas
Munck, Lars
Murkovic, Michael
Neicheva, Anastasia
Noteborn, Hubert
Nursten, Harry Erwin
Ochs, Stefan
Olano, Agustin
Olieman, Cornelis
Ollilainen, Velimatti
Ooghe, Wilfried
Österdahl, Bengt-Göran
Otte, Jeanette
Parr, Adrian James
Pavelka, Jiri
Pearce, Steven
Petz, Michael
Pfannhauser, Werner
Piironen, Vieno
Pischetsrieder, Monika
Plocková, Milada
Pokorny, Jan
Pokorný, Jan
Pospisil, Jasna
Poustka, Jan
Przysiezna, Ewa
Pudil, Frantisek
Rahali, Véronique

Restani, Patrizia
Ribarova, Fanny
Ridgway, Christopher
Rizov, Nicolay
Robak, Malgorzata
Rodziewicz, Anna
Roedig-Penman, Andrea
Rohn, Sascha
Rollin, Patrick
Rymowicz, Waldemar
Sakellariou, Christina
Salvo, Francesco
Sandberg, Ann-Sofie
Sass-Kiss, Agnes
Schäfer, Karola
Schlett, Claus
Schmid, Erich R.
Schmidt, Heinz
Schmidt, Stefan
Schreier, Peter
Schulte, Erhard
Schwenke, Klaus Dieter
Scordaki, Alexandra
Sforza, Stefano
Sheppard, Peter D.
Simko, Peter
Simon-Sarkadi, Livia
Sinigoj-Gacnik, Ksenija
Skrabka-Blotnicka, Teresa
Smith, Linda Bernhardine Margaret
Sontag, Gerhard
Speer, Karl
Steegmans, Monique
Steiner, Ingrid
Steinhart, Hans
Stetina, Jiri
Studer, Alfred
Surowka, Krzysztof
Szilágyi, Szilárd
Szoltysek, Katarzyna

Tateo, Fernando
Tauscher, Bernhard
Ternes, Waldemar
Thornton, Raymond E.
Timmermans, Eric
Topalova, Ivanka Chrstova
Uccella, Nicola Antonio
Ukmar Maljevac, Damjana
Ulberth, Franz
Vahteristo, Liisa
Van Boekel, Tiny
Van Den Bosch, Gerrit
Van Der Schee, Henk A.
Van Eckert, Renate
Van Peteghem, Carlos
Van Renterghem, Roland
Van Rhyn, Hans
Van Vyncht, Gery
Vanluchene, Eric
Velisek, Jan
Vidal-Valverde, Concepcion
Vieths, Stefan
Vischer, Michaela
Waldron, Keith
Wang, Rui
Weisz, Richard
Williams, John Graham
Williams, Mervyn
Williamson, Gary
Wilska-Jeszka, Jadwiga
Winkeler, Heinz-Dieter
Witkowska, Danuta
Zabetakis, Ioannis
Zegota, Henryk
Ziino, Marisa
Zoller, Otmar
Zoric, Andreja
Zunin, Paola
Zyla, Krzysztof

Hygiene

Anderssen, Valborg
Böhm, Josef
Budde, Jürgen
Chumchalová, Jana
Corvi, Claude Albert
Doganoc, Darinka Zdenka
Economides, Anna
Eklund, Trygve
Erning, Dieter
Galluser, Anita
Hauser, Eugen J. B.
Janson-Mundel, Ortrun
Jones, James
Keurulainen, Ritva
Kivistö, Laura
Kucera, Jiri

Lalljie, Sam
Lees, Ronald
Luf, Wolfgang
Moilanen, Raija
Nöhle, Ulrich
Northolt, Martin
Pezacki, Wincenty
Pipek, Petr
Plocková, Milada
Richter, Timo
Robak, Malgorzata
Rombouts, Frank
Rops, Wichard
Salvo, Francesco
Saukko, Maire
Schmidt, Heinz

Schmitt, Rudolf
Schmolck, Werner
Steiner, Ingrid
Stelz, Alice
Storgards, Erna
Suiraková, Eva
Thomas, Mark Andrew Llewellyn
Trojanowska, Krystyna
Tsanev, Roumen
Tuomala-Sarämaki, Terhi
Tykkyläinen, Paavo
Uijttenboogaart, Theo
Van Den Bosch, Gerrit
Vokk, Raivo
Whitehouse, Brian

Williams, Mervyn   Zagt, Robert   Ziajka, Stefan

Immunology

Benda, Vladimir
Hitchcock, Christopher
Kas, Jan
Klostermeyer, Henning

Korhonen, Hann
Mills, Elizabeth Naomi Clare
Morgan, Michael R.A.
Plaga-Lodde, Annette

Rauch, Pavel
Vieths, Stefan
Vodrazka, Zdenek

Irradiation

Bachman, Stefania
Basman, Arzu
Celik, Süeda
Chiavaro, Emma
Empis, José
Erning, Dieter

Farkas, Joszef
Goodman, Bernard
Hitchcock, Christopher
Ibe, Frank
Kas, Jan
Köksel, Hamit

Kontominas, Michael
Lalljie, Sam
Mörsel, Jörg-Thomas
Zegota, Henryk
Zoller, Otmar

Legumes

Alder, Lutz
Amarowicz, Ryszard
Basman, Arzu
Betsche, Thomas
Celik, Süeda
Coimbra, Manuel António
Czarnecka, Maria
Czarnecki, Zbgniew
Dalev, Pencho
Doncheva, Ivanka
Dostálová, Jana
Eshuis, Dolf F.
Fenwick, Gruffydd Roger
Frazier, Peter
Frias, Juana
Galluser, Anita
Györi, Zoltán
Hanewinkel-Meshkini, Susanne

Henderson, Nick C.
Henshall, David
Horvatic, Marija
Howard, Julie
Jones, Alan
Kerkvliet, Jacob
Khokhar, Santosh
Köksel, Hamit
Kostyra, Henryk
Kovacs, Elisabeth Teresia
Kozlowska, Halina
Krkoskova, Bernadetta
Lasztity, Radomir
Lindhauer, Meinolf G.
Morgan, Michael R.A.
Nowak, Jacek
Nursten, Harry Erwin
Pisciotta, Gennaro
Pospisil, Jasna

Rollin, Patrick
Sandberg, Ann-Sofie
Schwenke, Klaus Dieter
Sereno, Alberto M.C.
Sherlock, John C.
Siebers, Johannes
Sorensen, Susanne
Szilágyi, Szilárd
Tauscher, Bernhard
Thornton, Raymond E.
Tömösközi, Sándor
Vidal-Valverde, Concepcion
Waldron, Keith
Weder, Jürgen, K.P.
Wilska-Jeszka, Jadwiga
Winkler, Johanna
Yalcin, Erkan
Zyla, Krzysztof

Lipids and fatty acids

Adams, J. Brian
Ager, Elaine
Anklam, Elke
Aro, Tarja
Aubourg, Santiago
Balling Engelsen, Soren
Barron, Luis Javier Rodríguez
Beddows, Clifford G.
Beernaert, Hedwig
Beljaars, Paul
Böhm, Josef
Boskou, Dimitrius

Brathen, Gudmund
Brechany, Elizabeth
Buglass, Alan J.
Burini, Giovanni
Calzolari, Claudio
Christophersen, Carsten
Collar Esteve, Concha
Copíková, Jana
Damiani, Pietro
Demopolous, Constantinos A.
Deweghe, Liane
Di Luccia, Aldo

Dolezal, Marek
Dostálová, Jana
Ehlers, Dorothea
Ekar, Igor
Ellen, Geert
Estelecki, Ilona
Ferrara, Lydia
Filip, Vladimir
Fillery-Travis, Annette
Franzke, Claus
Gabrielli Favretto, Luciana
Gertz, Christian

Gojmerac, Tihomira
Gordon, Michael
Graille, Jean
Guillen, M. D.
Gunstone, Frank
Hardi, Jovica
Heinonen, Marina
Hermus, Rudolph J.J.
Hey, Michael James
Hitchcock, Christopher
Holmer, Gunhild
Honkavaara, Markku
Hopia, Anu
Jirovetz, Leopold
Jones, Alan
Juarez, Manuela
Kallio, Heikki
Kaufmann, Anton
Kedzior, Wladyslaw
Khokhar, Santosh
Knieling, Ralph G.
Komaitis, Michael
Kontominas, Michael
Kuusisto, Päivi
Laakso, Päivi
Lampi, Anna-Maija
Lautenbacher, Lutz-Michael
Le Botlan, Denis

Lees, Ronald
Lingnert, Hans
Löliger, J.
Lossonczy Von Losoncz, Thomas
Luckas, Bernd
Luf, Wolfgang
Luneia, Roberto
Mäkinen, Marjukka
Marini, Domenico
Marx, Friedhelm
Matissek, Reinhard
Mörsel, Jörg-Thomas
Mosandl, Armin
Moser, Ulrich
Murray, Brent Stuart
Nortvedt, Ragnar
Novakovic, Predrag
Pearce, J.
Piironen, Vieno
Pokorny, Jan
Pokorný, Jan
Primorac, Ljiljana
Ragotzky, Klaus
Ribarova, Fanny
Rihakova, Zdenka
Rizov, Nicolay
Rutkowski, Antoni

Rymowicz, Waldemar
Salvo, Francesco
Schmidt, Stefan
Schulte, Erhard
Schwack, Wolfgang
Schwarz, Walter
Sheppard, Peter D.
Skibsted, Leif H.
Smolinska, Teresa
Steinhart, Hans
Studer, Alfred
Szoltysek, Katarzyna
Tahvonen, Raija
Talbot, Geoff
Ternes, Waldemar
Thier, Hans-Peter
Toivo, Jari
Trojan, Marek
Tsanev, Roumen
Uccella, Nicola Antonio
Ulberth, Franz
Van Den Bosch, Gerrit
Vanluchene, Eric
Winkler, Johanna
Wood, Brian J.B.
Yanishlieva-Maslarova, Nedjalka
Ziino, Marisa

Maillard reaction

Ames, Jennifer
Bachman, Stefania
Baigrie, Brian
Baines, David Allan
Baltes, Werner
Birlouez, Inès
Botrè, Claudio
Cejpek, Karel
Corradini, Claudio
Davidek, Jiri
De Block, Jan
Di Luccia, Aldo
Evangelisti, Filippo
Farmer, Linda
Feinberg, Max

Gramshaw, J.W.
Henle, Thomas
Jägerstad, Margaretha
Jirovetz, Leopold
Kostyra, Henryk
Kroh, Lothar W.
Lees, Ronald
Leitner, Erich
Lingnert, Hans
Löliger, J.
Luf, Wolfgang
Martinez-Castro, Isabel
Mlotkiewicz, Jerzy
Mottram, Donald
Nursten, Harry Erwin

Obretonov, Tzvetan
Olano, Agustin
Parker, Jane
Pilizota, Vlasta
Pischetsrieder, Monika
Sikora, Marek
Steinhart, Hans
Tateo, Fernando
Van Boekel, Tiny
Velisek, Jan
Vlater, Vladimír
Williams, John Graham
Zegota, Henryk
Zunin, Paola

Management

Ahola, Maarit
Ahvenainen, Juha M.I.
Anderssen, Valborg
Anklam, Elke
Ashurst, Philip Roy
Battaglia, Reto
Beernaert, Hedwig
Beljaars, Paul
Bontenbal, Edwin

Brockmann, Anneliese
Clark, Michael
Clutton, David
Cooper, Julian
Cornelese, Johan
Davies, Robert John
De Kruif, Kees C.G.
Eberle, Mike
Eklund, Trygve

Fenwick, Gruffydd Roger
Finney, Graham
Flowerden, Mary
Galluser, Anita
Hauser, Eugen J. B.
Henderson, Nick C.
Hils, Arno K. A.
Hodgson, Ian
Holch, Klaus

Imbs, Boguslaw
Jones, Arthur David
Jones, James
Kadi, Andreas
Kellner, Vladimir
Klein, Erich
Kovac, Milan
Kramer, Jörg
Lalljie, Sam
Lambert, Michael
Langendam, Johannes
Lautenbacher, Lutz-Michael
Leloux, Mirjam

Lingnert, Hans
Malwitz, Dietmar
Martin, Peter Gerard
Nöhle, Ulrich
O´Brien, John
Österdahl, Bengt-Göran
Pipek, Petr
Preuss, Axel
Räsänen, Janne
Richter, Timo
Rizov, Nicolay
Roel, Peter
Schmolck, Werner

Schwarz, Walter
Studer, Alfred
Talbot, Geoff
Thomas, Mark Andrew
Llewellyn
Tiefenbacher, Karl
Toikkanen, Kari
Van Dael, Peter
Wiedner, Peter
Williams, Mervyn
Wittenschläger, Lutz

Marine toxins

Baars, Aalbert Jan
Botrè, Francesco
Groothuis, Dirk G.

Luckas, Bernd
Mazzei, Franco
Ruiter, Adriaan

Schlatter, Christian
Schlett, Claus

Meat

Aalbersberg, Willem Y.
Baines, David Allan
Balling Engelsen, Soren
Bauer, Friedrich
Bazulic, Davorin
Benda, Vladimir
Böhm, Josef
Bontenbal, Edwin
Botrè, Francesco
Brathen, Gudmund
Brockmann, Rainer
Brown, Peter Anthony
Brunn, Hubertus
Byrne, Briege Eileen
Campbell, Duncan J.
Cerkvenik, Vesna
Chiavaro, Emma
Culik, Jirí
Dalev, Pencho
Dawihl, Gerd
De Block, Jan
De Brabander, Hubert
De Jong, Jacob
De Ruyck, Hendrik
Di Luccia, Aldo
Doganoc, Darinka Zdenka
Dossena, Arnaldo
Economides, Anna
Enders, Peter W.
Estelecki, Ilona
Farkas, Joszef
Farkas, Pavel
Farmer, Linda
Finney, Graham
Galluser, Anita
Glück, Bernfried

Gojmerac, Tihomira
Groothuis, Dirk G.
Guillen, M. D.
Hajduk, Ewa
Hampton, Ian
Hauser, Eugen J. B.
Heinonen, Marina
Henderson, Nick C.
Herman, Lieve
Honikel, Karl Otto
Honkavaara, Markku
Hood, Ted D.E.
Ibe, Frank
Ilmoja, Kalle
Ingr, Ivo
Jägerstad, Margaretha
Jarmoluk, Andrzej
Jones, James
Kan, Kees (C.A.)
Kaufmann, Anton
Kedzior, Wladyslaw
Kivistö, Laura
Klein, Erich
Knieling, Ralph G.
Koch, Herbert
Kontominas, Michael
Kordic, Jasna
Krala, Lucjan
Kramer, Jörg
Krkoskova, Bernadetta
Kucera, Jiri
Kvasnicka, Frantisek
Lalljie, Sam
Leibetseder, Josef
Littmann-Nienstedt, Sigrid
Luckas, Bernd

Luf, Wolfgang
Lugasi, Andrea
Malwitz, Dietmar
Marchelli, Rosangela
Mc Donald, Mark
Melchior Larsen, Lone
Mikova, Kamila
Mlotkiewicz, Jerzy
Mottram, Donald
Mraz, Igor
Munck, Lars
Murkovic, Michael
Novakovic, Predrag
O' Neill, Ian
Ooghe, Wilfried
Österdahl, Bengt-Göran
Palka, Krystyna
Parker, Jane
Pavelka, Jiri
Petz, Michael
Pezacki, Wincenty
Pfannhauser, Werner
Pipek, Petr
Plaga-Lodde, Annette
Przysiezna, Ewa
Rimkus, Gerhard G.
Ristow, Reinhard
Rutkowski, Antoni
Sapunar-Postruznik, Jasenka
Sarpeid, Hans-Jacob
Schmidt, Heinz
Sforza, Stefano
Sherlock, John C.
Sikorski, Zdzistaw
Simko, Peter
Simon-Sarkadi, Livia

Sinigoj-Gacnik, Ksenija
Skibsted, Leif H.
Skrabka-Blotnicka, Teresa
Skrökki, Leila Anneli
Smolinska, Teresa
Sontag, Gerhard
Stark, Jacques
Steinhart, Hans
Thompson, Kenneth Clive

Tiusanen, Sirkka
Trojan, Marek
Troy, Declan J.
Uijttenboogaart, Theo
Ulberth, Franz
Valentová, Helena
Van Den Bosch, Gerrit
Van Peteghem, Carlos
Van Rhyn, Hans

Van Vyncht, Gery
Weisz, Richard
Wijngaards, Gerrit
Wilson, Reginald
Zidaric, Metka
Ziino, Marisa
Zyla, Krzysztof

Metabolism

Aura, Anna-Marija
Aust, Olivier
Balduck, Paul
Berger, Ralf Günter
Betsche, Thomas
Böhm, Josef
Eerikäinen, Tero
Eisenbrand, Gerhard
Faulds, Craig
Grenby, Trevor H.
Harris, Caroline A.
Havenaar, Robert
Henle, Thomas
Holst, Birgit

Kas, Jan
Kieffer, Felix
Kroon, Paul
Leibetseder, Josef
Morgan, Michael R.A.
Moser, Ulrich
Neicheva, Anastasia
Novakovic, Predrag
O´Brien, John
Pearce, J.
Rauch, Pavel
Ribarova, Fanny
Rimkus, Gerhard G.
Robak, Malgorzata

Rollin, Patrick
Russell, Wendy Roslyn
Saarinen, Niina
Schrenk, Dieter
Siezen, Roland J.
Sorensen, Hilmer
Stanoeva, Elena
Stempiewicz, Regina
Thornton, Raymond E.
Uijttenboogaart, Theo
Van Vyncht, Gery
Williamson, Gary
Witkowska, Danuta

Microbiology

Anderssen, Valborg
Ashurst, Philip Roy
Battaglia, Reto
Beljaars, Paul
Benda, Vladimir
Berger, Ralf Günter
Bontenbal, Edwin
Brathen, Gudmund
Budde, Jürgen
Calvo, Marta Maria
Chumchalová, Jana
Decaris, Bernard
Deweghe, Liane
Eklund, Trygve
Erning, Dieter
Evidente, Antonio
Farkas, Joszef
Faulds, Craig
Garncarek, Barbara
Groothuis, Dirk G.
Hauser, Eugen J. B.
Havenaar, Robert
Herman, Lieve
Hood, Ted D.E.
Ilsbroux, Ingrid

Jones, James
Keurulainen, Ritva
Kivistö, Laura
Klupsch, Robert N.
Koskenkorva, Anneli
Kramer, Jörg
Lautenbacher, Lutz-Michael
Lesniak, Wladyslaw
Matilainen, Katri
Melzoch, Karel
Mikova, Kamila
Moilanen, Raija
Northolt, Martin
Nowak, Jacek
Pastoriza, Laura
Payne, Nigel Kenneth
Pezacki, Wincenty
Pietkiewicz, Jerzy Jan
Pipek, Petr
Plaga-Lodde, Annette
Plocková, Milada
Preuss, Axel
Richter, Timo
Rihakova, Zdenka
Robak, Malgorzata

Rodziewicz, Anna
Rombouts, Frank
Rops, Wichard
Rychtera, Mojmír
Salminen, Seppo
Salminen, Seppo Jaakko
Saukko, Maire
Schmitt, Rudolf
Schreier, Peter
Sorhaug, Terje
Stark, Jacques
Steiner, Ingrid
Stelz, Alice
Stempiewicz, Regina
Suiraková, Eva
Thompson, Kenneth Clive
Trojanowska, Krystyna
Tuomala-Saramäki, Terhi
Tykkyläinen, Paavo
Wilson, Peter D.G.
Witkowska, Danuta
Wojtatowicz, Maria
Wood, Brian J.B.
Zagt, Robert
Zyla, Krzysztof

Minerals

Amaro Pinto, Rui Manuel
Beddows, Clifford G.
Beernaert, Hedwig
Betsche, Thomas
Bontenbal, Edwin
Curda, Ladislav
Deelstra, Hendrik
Deweghe, Liane
Ellen, Geert
Gaucheron, Frédéric
Györi, Zoltán
Havenaar, Robert
Horne, David S.
Ilmoja, Kalle
Juarez, Manuela
Khokhar, Santosh

Kieffer, Felix
Klapec, Tomislav
Kmiecik, Waldemar Andrzej
Lautenbacher, Lutz-Michael
Lisiewska, Zofia Barbara
Long, Alan
Mandic, Milena L.
Nortvedt, Ragnar
Pfannhauser, Werner
Popov, Dimitre
Rizov, Nicolay
Sandberg, Ann-Sofie
Sapunar-Postruznik, Jasenka
Schlegel-Zawadzka, Malgorzata
Schneider, Rüdiger

Sebecic, Blazenka
Sheppard, Peter D.
Skibniewska, Krystyna A.
Spiro, Michael
Stelz, Alice
Tahvonen, Raija
Timmermans, Eric
Trojan, Marek
Tykkyläinen, Paavo
Ugarcic-Hardi, Zaneta
Van Dokkum, Wim
Varey, Jane Elizabeth
Zyla, Krzysztof

Minimal processing

Aalbersberg, Willem Y.
Aura, Anna-Marija
Birlouez, Inès
Cheftel, Jean-Claude
Dietrich, Helmut
Empis, José
Eshuis, Dolf F.
Hauser, Eugen J. B.
Holst, Birgit
Koskenkorva, Anneli

Lasztity, Radomir
Luf, Wolfgang
Matuszek, Tadeusz
Mraz, Igor
Northolt, Martin
Pearce, J.
Pietkiewicz, Jerzy Jan
Pilizota, Vlasta
Pospisil, Jasna
Ragotzky, Klaus

Rombouts, Frank
Saukko, Maire
Sereno, Alberto M.C.
Steinbuch, Erwin
Subaric, Drago
Suutarinen, Marjaana
Tauscher, Bernhard
Troy, Declan J.

Mass spectroscopy

Ames, Jennifer
Anklam, Elke
Baigrie, Brian
Baltes, Werner
Beernaert, Hedwig
Beljaars, Paul
Berardo, Nicola
Berger, Ralf Günter
Betsche, Thomas
Böhm, Josef
Bosset, Jacques Olivier
Brechany, Elizabeth
Buglass, Alan J.
Bykowski, Piotr Jan
Christoph, Norbert
Cotte, Virginie
Crossy, Paul
Damiani, Pietro
De Brabander, Hubert
De Ruyck, Hendrik
Demopolous, Constantinos A.
Dettweiler, Gerd
Di Luccia, Aldo

Dietrich, Helmut
Dirinck, Patrick
Eisenbrand, Gerhard
Enders, Peter W.
Evidente, Antonio
Farkas, Pavel
Farmer, Linda
Finnegan, Derek
Gates, Leonard Michael
Gilbert, John
Gojmerac, Tihomira
Goodall, David M.
Gramshaw, J.W.
Guillen, M. D.
Hahn, Harald
Hakala, Mari
Häkkinen, Sari
Hardi, Jovica
Hauser, Eugen J. B.
Heinzler, Matthias
Henle, Thomas
Herraiz Tomico, Tomas
Hietaniemi, Veli

Holandová, Katerina
Holmer, Gunhild
Holst, Birgit
Honikel, Karl Otto
Howard, Julie
Huopalahti, Rainer
Ibe, Frank
Ilmoja, Kalle
Jirovetz, Leopold
Kallio, Heikki
Kaufmann, Anton
Knieling, Ralph G.
Komaitis, Michael
Kombal, Ralph
Kontominas, Michael
Kroh, Lothar W.
Laakso, Päivi
Lach, Günter
Lalljie, Sam
Lehtonen, Pekka
Leitner, Erich
Liddle, Peter
Luckas, Bernd

Luf, Wolfgang
Luneia, Roberto
Määttä, Kaisu
Malisch, Rainer
Martin, Gérard
Martinez-Castro, Isabel
Marx, Friedhelm
Mc Donald, Mark
Mischnick, Petra
Mlotkiewicz, Jerzy
Molnar-Perl, Ibolya
Mosandl, Armin
Mottram, Donald
Murkovic, Michael
Neicheva, Anastasia
Obretonov, Tzvetan
Ogaard Madsen, Jorgen
Ollilainen, Velimatti
Österdahl, Bengt-Göran

Parker, Jane
Petz, Michael
Pischetsrieder, Monika
Poustka, Jan
Pozderovic, Andrija
Pudil, Frantisek
Restani, Patrizia
Rimkus, Gerhard G.
Rizov, Nicolay
Rohn, Sascha
Rollin, Patrick
Schlett, Claus
Schmid, Erich R.
Schmidt, Heinz
Schreier, Peter
Schreyen, Luc
Sforza, Stefano
Siebers, Johannes
Siegmund, Barbara

Stanoeva, Elena
Steiner, Ingrid
Steinhart, Hans
Talou, Thierry
Tateo, Fernando
Tauscher, Bernhard
Ternes, Waldemar
Tóth-Markus, Marianna
Uccella, Nicola Antonio
Ulberth, Franz
Valentini, Giuseppa
Van Peteghem, Carlos
Van Rhyn, Hans
Van Vyncht, Gery
Velisek, Jan
Williams, Mervyn
Winkeler, Heinz-Dieter
Ziino, Marisa
Zoller, Otmar

Mycotoxins

Anklam, Elke
Baars, Aalbert Jan
Bachman, Stefania
Beljaars, Paul
Berardo, Nicola
Betsche, Thomas
Böhm, Josef
Brockmann, Rainer
Brown, Peter Anthony
Chiavaro, Emma
Chumchalová, Jana
Davies, Robert John
Doncheva, Ivanka
Dossena, Arnaldo
Ehlers, Dorothea
Evidente, Antonio
Gatti, Gian Carlo
Gilbert, John
Gojmerac, Tihomira
Hietaniemi, Veli
Holandová, Katerina

Huopalahti, Rainer
Ibe, Frank
Ilmoja, Kalle
Jörissen, Urban
Kan, Kees (C.A.)
Kombal, Ralph
Krska, Rudolf
Lach, Günter
Lasztity, Radomir
Lehtonen, Pekka
Leibetseder, Josef
Lindhauer, Meinolf G.
Lossonczy Von Losoncz, Thomas
Luckas, Bernd
Luf, Wolfgang
Marchelli, Rosangela
Melzoch, Karel
Morgan, Michael R.A.
Nehring, Ulrich P.
Northolt, Martin

Ochs, Stefan
Otteneder, Herbert
Plocková, Milada
Poustka, Jan
Rauch, Pavel
Restani, Patrizia
Salminen, Seppo
Schlatter, Christian
Schmolck, Werner
Seibel, Wilfried
Steiner, Ingrid
Stempiewicz, Regina
Szilágyi, Szilárd
Thornton, Raymond E.
Van Den Bosch, Gerrit
Van Peteghem, Carlos
Van Renterghem, Roland
Vanluchene, Eric
Whitehouse, Brian
Zegota, Henryk
Zoller, Otmar

Nutrition

Ahola, Maarit
Aura, Anna-Marija
Aust, Olivier
Balduck, Paul
Battaglia, Reto
Bazulic, Davorin
Beddows, Clifford G.
Beernaert, Hedwig
Beljaars, Paul
Betsche, Thomas
Birlouez, Inès
Böhm, Josef

Bontenbal, Edwin
Branch, Simon
Burini, Giovanni
Calvo, Marta Maria
Cejpek, Karel
Chumchalová, Jana
Clark, David
Cooper, Julian
Dalev, Pencho
Decaris, Bernard
Deelstra, Hendrik
Demopolous, Constantinos A.

Dolezal, Marek
Doncheva, Ivanka
Dostálová, Jana
Fenwick, Gruffydd Roger
Fillery-Travis, Annette
Fryer, John
Gaudard - De Weck, Daniele
Gladovic, Natasa
Gordon, Michael
Grenby, Trevor Hilary
Gunstone, Frank
Häkkinen, Sari

Havenaar, Robert
Henle, Thomas
Hermus, Rudolph J.J.
Hodgson, Ian
Holasova, Marie
Holmer, Gunhild
Honkavaara, Markku
Horne-Ekman, Maarit
Horváth-Mosonyi, Magda
Horvatic, Marija
Hrncirik, Karel
Hruskar, Mirjana
Ingr, Ivo
Jirovetz, Leopold
Jones, Alan
Kadi, Andreas
Kallio, Heikki
Kedzior, Wladyslaw
Kerkvliet, Jacob
Keurulainen, Ritva
Khokhar, Santosh
Kieffer, Felix
Klaarenbeek, Tineke
Klapec, Tomislav
Knowles, Michael Ernest
Kordic, Jasna
Kosicki, Zenon
Koskenkorva, Anneli
Kovac, Spomenka
Kovacs, Elisabeth Teresia
Kozlowska, Halina
Kroll, Jürgen
Kroyer, Gerhard Th.
Langendam, Johannes

Lasztity, Radomir
Leibetseder, Josef
Leigh, Anthony
Long, Alan
Lossonczy Von Losoncz, Thomas
Luf, Wolfgang
Lugasi, Andrea
Luneia, Roberto
Mandic, Milena L.
Masková, Eva
Melzoch, Karel
Mikova, Kamila
Mikuschka, Gerhard
Miskiewicz, Tadeusz
Moser, Ulrich
Niemi, Sanna-Maria
Novakovic, Predrag
O´Brien, John
Ooghe, Wilfried
Pearce, J.
Pezacki, Wincenty
Pisciotta, Gennaro
Pokorny, Jan
Pokorný, Jan
Pollmer, Udo
Popken, Anne M.
Pospisil, Jasna
Primorac, Ljiljana
Prugar, Jaroslav
Rahali, Véronique
Ribarova, Fanny
Rizov, Nicolay
Roedig-Penman, Andrea

Ruiter, Adriaan
Russell, Wendy Roslyn
Salminen, Seppo
Sandberg, Ann-Sofie
Sapunar-Postruznik, Jasenka
Schlegel-Zawadzka, Malgorzata
Schrenk, Dieter
Sebecic, Blazenka
Seibel, Wilfried
Sheppard, Peter D.
Stelz, Alice
Suhaj, Milan
Tahvonen, Raija
Thornton, Raymond E.
Timmermans, Eric
Tömösközi, Sándor
Tsanev, Roumen
Ukmar Maljevac, Damjana
Vahteristo, Liisa
Van Dael, Peter
Van Den Bosch, Gerrit
Van Dokkum, Wim
Van Peteghem, Carlos
Varey, Jane Elizabeth
Vedrina-Dragojevic, Irena
Vidal-Valverde, Concepcion
Vokk, Raivo
Weder, Jürgen, K.P.
Williams, John Graham
Williamson, Gary
Wilson, Peter D.G.
Zagt, Robert
Zyla, Krzysztof

Oils and fats

Adamantiadou, Sophia
Ager, Elaine
Anklam, Elke
Balling Engelsen, Soren
Barron, Luis Javier Rodríguez
Beddows, Clifford G.
Beljaars, Paul
Boast, Martin
Böhm, Josef
Boskou, Dimitrius
Brathen, Gudmund
Campanella, Luigi
Clark, Michael
Damiani, Pietro
De Jong, Jacob
De Ruiter, Gerhard A.
Demopolous, Constantinos A.
Di Natale, Corrado
Dirinck, Patrick
Dolezal, Marek
Ehlers, Dorothea
Ekar, Igor
Empis, José

Estelecki, Ilona
Evangelisti, Filippo
Favretto, Luciano
Filip, Vladimir
Fillery-Travis, Annette
Finney, Graham
Fisher, Leonard
Franzke, Claus
Gabrielli Favretto, Luciana
Gertz, Christian
Gordon, Michael
Graille, Jean
Guillen, M. D.
Gunstone, Frank
Heinonen, Marina
Heredia, Antonia
Hey, Michael James
Hills, Brian
Hitchcock, Christopher
Holandová, Katerina
Holmer, Gunhild
Hopia, Anu
Ilsbroux, Ingrid

Imbs, Boguslaw
Ivanov, Kalintcho
Jörissen, Urban
Juarez, Manuela
Kallio, Heikki
Klaarenbeek, Tineke
Komaitis, Michael
Kuusisto, Päivi
Laakso, Päivi
Lach, Günter
Lampi, Anna-Maija
Le Botlan, Denis
Löliger, J.
Lossonczy Von Losoncz, Thomas
Luckas, Bernd
Luf, Wolfgang
Luneia, Roberto
Mäkinen, Marjukka
Malwitz, Dietmar
Marini, Domenico
Marx, Friedhelm
Mörsel, Jörg-Thomas

Mosandl, Armin
Murkovic, Michael
Murray, Brent Stuart
Nöhle, Ulrich
Pearce, J.
Pfannhauser, Werner
Pisciotta, Gennaro
Plocková, Milada
Pokorny, Jan
Pokorný, Jan
Ragotzky, Klaus
Rihakova, Zdenka
Rutkowski, Antoni
Salvo, Francesco
Schmidt, Stefan

Schulte, Erhard
Schutte, Leonard
Schwack, Wolfgang
Schwarz, Walter
Scordaki, Alexandra
Skrökki, Leila Anneli
Studer, Alfred
Suhaj, Milan
Szoltysek, Katarzyna
Talbot, Geoff
Ternes, Waldemar
Thornton, Raymond E.
Tóth-Markus, Marianna
Tsanev, Roumen
Uccella, Nicola Antonio

Ulberth, Franz
Valentová, Helena
Van Den Bosch, Gerrit
Van Vyncht, Gery
Venskutonis, Petras Rimantas
Weisz, Richard
Wilson, Reginald
Winkler, Johanna
Yanishlieva-Maslarova,
 Nedjalka
Zidaric, Metka
Ziino, Marisa
Zunin, Paola

Oxidation

Aubourg, Santiago
Beddows, Clifford G.
Birlouez, Inès
Bontempelli, Gino
Christophersen, Carsten
Damiani, Pietro
Davies, Alan Philipp
Empis, José
Eshuis, Dolf F.
Evangelisti, Filippo
Farkas, Joszef
Farmer, Linda
Filip, Vladimir
Finnegan, Derek
Gertz, Christian

Goodman, Bernard
Henle, Thomas
Holmer, Gunhild
Hopia, Anu
Lampi, Anna-Maija
Lingnert, Hans
Luf, Wolfgang
Lugasi, Andrea
Mäkinen, Marjukka
Makris, Dimitris
Melchior Larsen, Lone
Murkovic, Michael
O' Neill, Ian
Piironen, Vieno
Pokorny, Jan

Rollin, Patrick
Russell, Wendy Roslyn
Skibsted, Leif H.
Spiro, Michael
Tauscher, Bernhard
Thornton, Raymond E.
Tsanev, Roumen
Von Wietersheim, Eugen
Waldron, Keith
Wilska-Jeszka, Jadwiga
Yanishlieva-Maslarova,
 Nedjalka
Zegota, Henryk
Zunin, Paola

Packaging

Anklam, Elke
Ashurst, Philip Roy
Aura, Anna-Marija
Bosset, Jacques Olivier
Catalá, Ramon
Dirinck, Patrick
Eberle, Mike
Galic, Kata
Gatti, Gian Carlo
Gilbert, John
Golja, Viviana
Hasenay, Damir
Hauser, Eugen J. B.
Hood, Ted D.E.

Howick, Chris
Imbs, Boguslaw
Järvi-Käärinäinen, Irma Terhen
Jones, James
Knowles, Michael Ernest
Kontominas, Michael
Krala, Lucjan
Lalljie, Sam
Leitner, Erich
Matuszek, Tadeusz
Mraz, Igor
Pastoriza, Laura
Pezacki, Wincenty
Pilizota, Vlasta

Pipek, Petr
Pospisil, Jasna
Schmolck, Werner
Simko, Peter
Steiner, Ingrid
Talou, Thierry
Tauscher, Bernhard
Toikkanen, Kari
Uijttenboogaart, Theo
Von Wietersheim, Eugen
Williams, Mervyn
Wilson, Peter D.G.
Wittenschläger, Lutz
Zoric, Andreja

PAHs

Baars, Aalbert Jan
Baigrie, Brian
Beernaert, Hedwig

Botrè, Claudio
Bykowski, Piotr Jan
Christoph, Norbert

Crossy, Paul
Culik, Jirí
Eisenbrand, Gerhard

Gertz, Christian
Glück, Bernfried
Graille, Jean
Guillen, M. D.
Hahn, Harald
Hietaniemi, Veli
Honikel, Karl Otto
Huopalahti, Rainer
Ilmoja, Kalle
Kellner, Vladimir
Klein, Erich
Koch, Herbert
Kombal, Ralph

Krska, Rudolf
Lach, Günter
Lalljie, Sam
Melzoch, Karel
Moret, Ivo
Pfalzgraf, Andreas
Pfannhauser, Werner
Pipek, Petr
Poustka, Jan
Rantamäkki, Pirjio
Rizov, Nicolay
Schlatter, Christian
Schlett, Claus

Schmid, Erich R.
Schrenk, Dieter
Simko, Peter
Speer, Karl
Thornton, Raymond E.
Tiefenbacher, Karl
Van Vyncht, Gery
Wang, Rui
Weisz, Richard
Wiedner, Peter
Winkeler, Heinz-Dieter

PCBs

Anklam, Elke
Baars, Aalbert Jan
Baigrie, Brian
Bauer, Ulrich
Bazulic, Davorin
Beernaert, Hedwig
Betsche, Thomas
Botrè, Claudio
Brockmann, Rainer
Brunn, Hubertus
Bykowski, Piotr Jan
Corvi, Claude Albert
Crossy, Paul
Culik, Jirí
Daphi-Weber, Juliane
De Brabander, Hubert
Gatti, Gian Carlo

Hahn, Harald
Harris, Caroline A.
Hietaniemi, Veli
Holandová, Katerina
Honikel, Karl Otto
Ilmoja, Kalle
Kan, Kees (C.A.)
Kellner, Vladimir
Koch, Herbert
Lach, Günter
Lalljie, Sam
Luckas, Bernd
Malisch, Rainer
Moret, Ivo
Pfalzgraf, Andreas
Poustka, Jan
Rimkus, Gerhard G.

Rizov, Nicolay
Rops, Wichard
Ruiter, Adriaan
Schlatter, Christian
Schlett, Claus
Schmid, Erich R.
Schrenk, Dieter
Schreyen, Luc
Ternes, Waldemar
Thier, Hans-Peter
Thornton, Raymond E.
Van Peteghem, Carlos
Van Renterghem, Roland
Van Vyncht, Gery
Vanluchene, Eric
Winkeler, Heinz-Dieter
Zidaric, Metka

PCR

Amaro Pinto, Rui Manuel
Anklam, Elke
Beljaars, Paul
Berger, Ralf Günter
Chumchalová, Jana
Decaris, Bernard
Eisenbrand, Gerhard
Glück, Bernfried

Henle, Thomas
Honikel, Karl Otto
Krause, Wolfgang
Luf, Wolfgang
Martelli, Aldo
Melzoch, Karel
Meunier, Jean-Claude
Plocková, Milada

Schlett, Claus
Schmolck, Werner
Suiraková, Eva
Thornton, Raymond E.
Varadi, Maria
Vieths, Stefan
Weder, Jürgen, K.P.

Pesticides

Alder, Lutz
Anklam, Elke
Baars, Aalbert Jan
Baigrie, Brian
Bauer, Ulrich
Bazulic, Davorin
Beernaert, Hedwig
Benda, Vladimir
Betsche, Thomas

Brunn, Hubertus
Buglass, Alan J.
Bykowski, Piotr Jan
Campanella, Luigi
Cerkvenik, Vesna
Corvi, Claude Albert
Crossy, Paul
Damiani, Pietro
Dettweiler, Gerd

Enders, Peter W.
Finnegan, Derek
Gatti, Gian Carlo
Gojmerac, Tihomira
Hardi, Jovica
Harris, Caroline A.
Heinzler, Matthias
Holandová, Katerina
Ibe, Frank

Ilmoja, Kalle
Kan, Kees (C.A.)
Knieling, Ralph G.
Kombal, Ralph
Kovac, Spomenka
Krska, Rudolf
Lach, Günter
Lalljie, Sam
Leitner, Erich
Luckas, Bernd
Malisch, Rainer
Mazzei, Franco
Moret, Ivo
Morgan, Michael R.A.
Nehring, Ulrich P.
Neicheva, Anastasia
O´Brien, John

Ochs, Stefan
Österdahl, Bengt-Göran
Otteneder, Herbert
Pfalzgraf, Andreas
Pfannhauser, Werner
Pollmer, Udo
Poustka, Jan
Rauch, Pavel
Restani, Patrizia
Rimkus, Gerhard G.
Ristow, Reinhard
Rizov, Nicolay
Salvo, Francesco
Schlatter, Christian
Schlett, Claus
Schmid, Erich R.
Schrenk, Dieter

Schwack, Wolfgang
Seibel, Wilfried
Sherlock, John C.
Siebers, Johannes
Speer, Karl
Stanoeva, Elena
Tennant, David
Thier, Hans-Peter
Topalova, Ivanka Chrstova
Van Den Bosch, Gerrit
Van Der Schee, Henk A.
Van Vyncht, Gery
Vanluchene, Eric
Vischer, Michaela
Winkeler, Heinz-Dieter
Zidaric, Metka

Plant toxins

Anklam, Elke
Baars, Aalbert Jan
Betsche, Thomas
Böhm, Josef
Dalev, Pencho
Davidek, Jiri
Dos Santos Baptista, Bráulio
Enders, Peter W.
Fenwick, Gruffydd Roger
Frias, Juana
Hermus, Rudolph J.J.
Holst, Birgit

Hrncirik, Karel
Huopalahti, Rainer
Ilmoja, Kalle
Kalac, Pavel
Knieling, Ralph G.
Knowles, Michael Ernest
Kucera, Jiri
Lalljie, Sam
Morgan, Michael R.A.
Noteborn, Hubert
O' Neill, Ian
O´Brien, John

Parr, Adrian James
Pollmer, Udo
Rollin, Patrick
Schlatter, Christian
Sherlock, John C.
Stanoeva, Elena
Suhaj, Milan
Van Vyncht, Gery
Weder, Jürgen, K.P.
Zoller, Otmar

Polarography

Beljaars, Paul
Ellen, Geert

Hasenay, Damir
Ilmoja, Kalle

Knieling, Ralph G.
Ternes, Waldemar

Polyphenols

Adams, J. Brian
Amarowicz, Ryszard
Anklam, Elke
Aura, Anna-Marija
Beddows, Clifford G.
Böhm, Volker
Boskou, Dimitrius
Botrè, Francesco
Campanella, Luigi
Christophersen, Carsten
Chrysafidis, Dimitrios
Davies, Alan Philip
Dietrich, Helmut
Eshuis, Dolf F.
Evangelisti, Filippo
Evidente, Antonio

Faulds, Craig
Gonzalez-Sanjose, Maria Luisa
Gordon, Michael
Graille, Jean
Gramshaw, J.W.
Häkkinen, Sari
Heinonen, Marina
Henle, Thomas
Heredia, Antonia
Hopia, Anu
Jones, Arthur David
Khokhar, Santosh
Knowles, Michael Ernest
Korhonen, Hann
Kozlowska, Halina
Kroll, Jürgen

Kroon, Paul
Kroyer, Gerhard Th.
Kvasnicka, Frantisek
Lehtonen, Pekka
Long, Alan
Luf, Wolfgang
Lugasi, Andrea
Makris, Dimitris
Martelli, Aldo
Melchior Larsen, Lone
Melzoch, Karel
Noteborn, Hubert
Nursten, Harry Erwin
Ooghe, Wilfried
Parr, Adrian James
Pearce, J.

Pfannhauser, Werner
Pilizota, Vlasta
Pokorný, Jan
Popken, Anne M.
Ragotzky, Klaus
Roedig-Penman, Andrea
Rohn, Sascha
Russell, Wendy Roslyn
Saarinen, Niina

Saltmarsh, Mike
Salvo, Francesco
Schmidt, Stefan
Sime, John
Smith, Linda Bernhardine Margaret
Spiro, Michael
Studer, Alfred
Subaric, Drago

Ternes, Waldemar
Thornton, Raymond E.
Uccella, Nicola Antonio
Waldron, Keith
Williamson, Gary
Wilska-Jeszka, Jadwiga
Yanishlieva-Maslarova, Nedjalka
Zunin, Paola

Potatoes

Aalbersberg, Willem Y.
Adams, J. Brian
Ames, Jennifer
Bachman, Stefania
Betsche, Thomas
Budde, Jürgen
Büning-Pfaue, Hans
Czarnecka, Maria
Czarnecki, Zbgniew
Donald, Athene
Eberle, Mike
Eshuis, Dolf F.
Galluser, Anita
Golachowski, Antoni

Heinonen, Marina
Heinzler, Matthias
Huopalahti, Rainer
Ilmoja, Kalle
Jones, James
Kerkvliet, Jacob
Klaarenbeek, Tineke
Leszczynski, Waclaw
Lindhauer, Meinolf G.
Lisinska, Grazyna
Marioleas, Panagiotis
Melchior Larsen, Lone
Miskiewicz, Tadeusz
Mraz, Igor

Murkovic, Michael
Nehring, Ulrich P.
Palasinski, Mieczyslaw
Panovská, Zdenka
Peksa, Anna
Pozderovic, Andrija
Prugar, Jaroslav
Seibel, Wilfried
Sherlock, John C.
Szilágyi, Szilárd
Vella, Anthony
Waldron, Keith

Preservation

Bachman, Stefania
Beljaars, Paul
Chumchalová, Jana
Czarnecka, Maria
Di Luccia, Aldo
Drdak, Milan
Eberle, Mike
Eklund, Trygve
Eshuis, Dolf F.
Farkas, Joszef
Guillen, M. D.
Henshall, David
Iciek, Jan
Ingr, Ivo
Jarmoluk, Andrzej
Jirovetz, Leopold

Kvasnicka, Frantisek
Lalljie, Sam
Marin, Michèle
Mikova, Kamila
Mraz, Igor
Nehring, Ulrich P.
Northolt, Martin
Palka, Krystyna
Pastoriza, Laura
Pezacki, Wincenty
Pilizota, Vlasta
Pipek, Petr
Plocková, Milada
Przysiezna, Ewa
Ridgway, Christopher
Rihakova, Zdenka

Roel, Peter
Rombouts, Frank
Rutkowski, Antoni
Sikorski, Zdzistaw
Skrabka-Blotnicka, Teresa
Stark, Jacques
Steinbuch, Erwin
Steiner, Ingrid
Suiraková, Eva
Tykkyläinen, Paavo
Uijttenboogaart, Theo
Venskutonis, Petras Rimantas
Wojtatowicz, Maria
Wucherpfennig, Karl
Zegota, Henryk

Proteins

Aalbersberg, Willem Y.
Adams, J. Brian
Amarowicz, Ryszard
Andrews, Anthony
Aust, Olivier
Basman, Arzu
Bauer, Friedrich
Beljaars, Paul

Belton, Peter
Benda, Vladimir
Berardo, Nicola
Betsche, Thomas
Birlouez, Inès
Brathen, Gudmund
Calvo, Marta Maria
Celik, Süeda

Cheftel, Jean-Claude
Chroneos, Ioannis
Clark, David
Collar Esteve, Concha
Curda, Ladislav
Dalev, Pencho
Dawihl, Gerd
De Block, Jan

De Kruif, Kees C.G.
De Ruiter, Gerhard A.
Delgadillo, Ivonne
Deweghe, Liane
Di Luccia, Aldo
Dickinson, Eric
Dziuba, Jerzy
Dzwolak, Waldemar
Eisenbrand, Gerhard
Estelecki, Ilona
Evidente, Antonio
Fillery-Travis, Annette
Fisher, Leonard
Frazier, Peter
Gaucheron, Frédéric
Gil, Ana Maria Pissarra C.
Glück, Bernfried
Gojmerac, Tihomira
Hajduk, Ewa
Havenaar, Robert
Henderson, Nick C.
Henle, Thomas
Horne, David S.
Horvatic, Marija
Ilari, Jean-Luc
Jones, Arthur David
Kas, Jan
Klostermeyer, Henning
Köksel, Hamit
Korhonen, Hann
Kostyra, Henryk
Krala, Lucjan
Krause, Wolfgang
Kroll, Jürgen
Kroon, Paul

Kucera, Jiri
Lalljie, Sam
Lapveteläinen, Anja
Lasztity, Radomir
Lindhauer, Meinolf G.
Lisinska, Grazyna
Long, Alan
Ludwig, Eberhard
Luf, Wolfgang
Marchelli, Rosangela
Mills, Elizabeth Naomi Clare
Morris, Victor John
Murray, Brent Stuart
Noteborn, Hubert
Olieman, Cornelis
Ooghe, Wilfried
Otte, Jeanette
Palka, Krystyna
Peksa, Anna
Pezacki, Wincenty
Pipek, Petr
Przysiezna, Ewa
Rahali, Véronique
Rantamäkki, Pirjio
Restani, Patrizia
Rizov, Nicolay
Robak, Malgorzata
Rodziewicz, Anna
Rohn, Sascha
Rutkowski, Antoni
Salgò, Andràs
Sarpeid, Hans-Jacob
Sass-Kiss, Agnes
Schäfer, Karola
Schwenke, Klaus Dieter

Sebecic, Blazenka
Sforza, Stefano
Siezen, Roland J.
Sikorski, Zdzistaw
Simon-Sarkadi, Livia
Skrabka-Blotnicka, Teresa
Sorensen, Hilmer
Sorensen, Susanne
Sorhaug, Terje
Sousa, Isabel
Stetina, Jiri
Surowka, Krzysztof
Szilágyi, Szilárd
Tauscher, Bernhard
Tiefenbacher, Karl
Timmermans, Eric
Tömösközi, Sándor
Troy, Declan J.
Uijttenboogaart, Theo
Van Boekel, Tiny
Van Den Bosch, Gerrit
Van Eckert, Renate
Van Vyncht, Gery
Vieths, Stefan
Voragen, Fons
Waldron, Keith
Weder, Jürgen, K.P.
Wijngaards, Gerrit
Wilde, Peter
Wilson, Philip
Winkler, Johanna
Witkowska, Danuta
Yalcin, Erkan
Ziino, Marisa
Zyla, Krzysztof

Quality

Anderssen, Valborg
Anklam, Elke
Assimakopoulou, Angelique
Aubourg, Santiago
Baines, David Allan
Battaglia, Reto
Bauer, Friedrich
Beernaert, Hedwig
Benedito, Carmen
Berardo, Nicola
Berger, Ralf Günter
Betsche, Thomas
Birlouez, Inès
Bjergegaard, Charlotte
Boast, Martin
Bubnik, Zdenek
Calzolari, Claudio
Campbell, Duncan J.
Clutton, David
Corradini, Claudio
Coveney, Leslie
Curda, Ladislav

Damiani, Pietro
De Ruyck, Hendrik
Delgadillo, Ivonne
Di Luccia, Aldo
Doganoc, Darinka Zdenka
Donald, Athene
Dostálová, Jana
Drdak, Milan
Ducauze, Christian J.
Duran, Luis
Dzwolak, Waldemar
Eerikäinen, Tero
Ellen, Geert
Enders, Peter W.
Ennion, Ronald
Feinberg, Max
Finnegan, Derek
Finney, Graham
Flowerden, Mary
Galluser, Anita
Gates, Leonard Michael
Gatti, Gian Carlo

Gaucheron, Frédéric
Gertz, Christian
Gladovic, Natasa
Glück, Bernfried
Gonzalez-Sanjose, Maria Luisa
Goodman, Bernard
Guillen, M. D.
Hägg, Margareta
Hamilton, Colin A.
Hauser, Eugen J. B.
Havenaar, Robert
Henle, Thomas
Herman, Lieve
Hills, Brian
Hils, Arno K. A.
Hitchcock, Christopher
Hood, Ted D.E.
Horne, David S.
Hruskar, Mirjana
Iciek, Jan
Ilmoja, Kalle
Imbs, Boguslaw

163

Ingr, Ivo
Jarmoluk, Andrzej
Jones, James
Kadi, Andreas
Kedzior, Wladyslaw
Keurulainen, Ritva
Kivistö, Laura
Klein, Erich
Knieling, Ralph G.
Kontominas, Michael
Krala, Lucjan
Kramer, Jörg
Kuusisto, Päivi
Lalljie, Sam
Lambert, Michael
Lasztity, Radomir
Lees, Ronald
Lingnert, Hans
Lisiewska, Zofia Barbara
Lisinska, Grazyna
Lorenzen, Kay
Lossonczy Von Losoncz, Thomas
Luf, Wolfgang
Mac Dowall, James
Maclean, Wim
Malwitz, Dietmar
Marchelli, Rosangela
Martin, Peter Gerard
Martinez Anaya, Antonia M.
Matilainen, Katri
Mc Donald, Mark
Melzoch, Karel

Mikova, Kamila
Mikuschka, Gerhard
Moilanen, Raija
Molnar, Pal
Mörsel, Jörg-Thomas
Mraz, Igor
Murray, Brent Stuart
Nöhle, Ulrich
Nortvedt, Ragnar
Otte, Jeanette
Parr, Adrian James
Payne, Nigel Kenneth
Pezacki, Wincenty
Pfalzgraf, Andreas
Pietkiewicz, Jerzy Jan
Pipek, Petr
Pisciotta, Gennaro
Plocková, Milada
Pollmer, Udo
Preuss, Axel
Prugar, Jaroslav
Pudil, Frantisek
Rahali, Véronique
Ridgway, Christopher
Rizov, Nicolay
Rutkowski, Antoni
Rutledge, Douglas
Salvo, Francesco
Schmidt, Stefan
Schmolck, Werner
Schrenk, Dieter
Schreyen, Luc
Schwarz, Walter

Sikorski, Zdzistaw
Sinková, Terèzia
Skibniewska, Krystyna A.
Skibsted, Leif H.
Smith, Linda Bernhardine Margaret
Smolinska, Teresa
Steinbuch, Erwin
Steiner, Ingrid
Stetina, Jiri
Suhaj, Milan
Szilágyi, Szilárd
Szoltysek, Katarzyna
Talou, Thierry
Tauscher, Bernhard
Thomas, Mark Andrew
Llewellyn
Troy, Declan J.
Tuomala-Saramäki, Terhi
Ugarcic-Hardi, Zaneta
Uijttenboogaart, Theo
Ukmar Maljevac, Damjana
Vahcic, Nada
Van Rhyn, Hans
Varadi, Maria
Varadi, Maria
Vischer, Michaela
Von Wietersheim, Eugen
Wiggins, Edgar Hugh
Wilde, Peter
Williams, Mervyn
Wilson, Philip
Ziajka, Stefan

Quality assurance

Anderssen, Valborg
Baines, David Allan
Bazulic, Davorin
Beernaert, Hedwig
Beljaars, Paul
Boast, Martin
Bykowski, Piotr Jan
Calzolari, Claudio
Clutton, David
Cotte, Virginie
Coveney, Leslie
Doncheva, Ivanka
Ducauze, Christian J.
Dzwolak, Waldemar
Eberle, Mike
Economides, Anna
Ellen, Geert
Enders, Peter W.
Erning, Dieter
Filip, Vladimir
Finnegan, Derek
Finney, Graham
Flowerden, Mary
Galluser, Anita

Gates, Leonard Michael
Gatti, Gian Carlo
Gertz, Christian
Gilbert, John
Gladovic, Natasa
Glück, Bernfried
Gojmerac, Tihomira
Groothuis, Dirk G.
Györi, Zoltán
Hägg, Margareta
Hamilton, Colin A.
Hauser, Eugen J. B.
Herman, Lieve
Hey, Hanke
Hruskar, Mirjana
Ilmoja, Kalle
Janson-Mundel, Ortrun
Jones, James
Kadi, Andreas
Kedzior, Wladyslaw
Kivistö, Laura
Klaarenbeek, Tineke
Klein, Bernard
Klein, Erich

Knieling, Ralph G.
Koskenkorva, Anneli
Kovac, Milan
Kramer, Jörg
Kuusisto, Päivi
Lalljie, Sam
Lambert, Michael
Lorenzen, Kay
Lossonczy Von Losoncz, Thomas
Luf, Wolfgang
Maclean, Wim
Malwitz, Dietmar
Marini, Domenico
Martelli, Aldo
Melzoch, Karel
Mikova, Kamila
Mikuschka, Gerhard
Moilanen, Raija
Molnar, Pal
Mosandl, Armin
Mraz, Igor
Nöhle, Ulrich
Northolt, Martin

Ollilainen, Velimatti
Pearce, Steven
Pezacki, Wincenty
Pipek, Petr
Pisciotta, Gennaro
Plocková, Milada
Rizov, Nicolay
Rombouts, Frank
Rutledge, Douglas
Saltmarsh, Mike
Saukko, Maire
Schmitt, Rudolf
Schmolck, Werner
Schreyen, Luc
Schwarz, Walter

Sheppard, Peter D.
Steinbuch, Erwin
Steiner, Ingrid
Storgards, Erna
Suhaj, Milan
Szilágyi, Szilárd
Szoltysek, Katarzyna
Thier, Hans-Peter
Thomas, Mark Andrew
Llewellyn
Tiefenbacher, Karl
Tuomala-Saramäki, Terhi
Ukmar Maljevac, Damjana
Vahcic, Nada
Van Den Bosch, Gerrit

Van Rhyn, Hans
Varadi, Maria
Varadi, Maria
Vella, Anthony
Venskutonis, Petras Rimantas
Vischer, Michaela
Von Wietersheim, Eugen
Weisz, Richard
Whitehouse, Brian
Williams, Mervyn
Wittenschläger, Lutz
Zagt, Robert
Ziajka, Stefan

Regulative issues

Adamantiadou, Sophia
Anklam, Elke
Assimakopoulou, Angelique
Battaglia, Reto
Beernaert, Hedwig
Betsche, Thomas
Boskou, Dimitrius
Campbell, Duncan J.
Clutton, David
Davies, Robert John
De Jong, Jacob
De Ruyck, Hendrik
Ennion, Ronald
Erning, Dieter
Finnegan, Derek
Finney, Graham
Gilbert, John
Golja, Viviana
Gramshaw, J.W.
Grenby, Trevor Hilary
Hampton, Ian
Harris, Caroline A.
Henshall, David

Hitchcock, Christopher
Hodgson, Ian
Howick, Chris
Jones, James
Kadi, Andreas
Keurulainen, Ritva
Klaarenbeek, Tineke
Klein, Bernard
Knieling, Ralph G.
Kombal, Ralph
Lalljie, Sam
Lambert, Michael
Landsiedel, Robert
Lautenbacher, Lutz-Michael
Liddle, Peter
Malwitz, Dietmar
Martin, Peter Gerard
Mc Donald, Mark
Mikuschka, Gerhard
Molnar, Pal
Nehring, Ulrich P.
Nöhle, Ulrich
O´Brien, John

Payne, Nigel Kenneth
Pfalzgraf, Andreas
Preuss, Axel
Ristow, Reinhard
Rizov, Nicolay
Rohn, Sascha
Saltmarsh, Mike
Schrenk, Dieter
Scordaki, Alexandra
Sheppard, Peter D.
Siebers, Johannes
Stanoeva, Elena
Temmerman, Guy
Tennant, David
Thornton, Raymond E.
Vahcic, Nada
Van Rhyn, Hans
Von Rymon Lipinski, Gert-Wolfhard
Whitehouse, Brian
Wiggins, Edgar Hugh
Wittenschläger, Lutz
Zoric, Andreja

Rheology

Balduck, Paul
Basman, Arzu
Benedito, Carmen
Betsche, Thomas
Celik, Süeda
Clark, David
Clark, Michael
Collar Esteve, Concha
Curda, Ladislav
De Kruif, Kees C.G.
De Ruiter, Gerhard A.
Dickinson, Eric
Donald, Athene
Duran, Luis

Dziuba, Jerzy
Estelecki, Ilona
Filip, Vladimir
Fillery-Travis, Annette
Fisher, Leonard
Frazier, Peter
Garncarek, Barbara
Hegedusic, Vesna
Henle, Thomas
Hodgson, Ian
Horne, David S.
Ilari, Jean-Luc
Köksel, Hamit
Martinez Anaya, Antonia M.

Matuszek, Tadeusz
Morris, Victor John
Murray, Brent Stuart
Palasinski, Mieczyslaw
Palka, Krystyna
Panovská, Zdenka
Peksa, Anna
Pezacki, Wincenty
Pisciotta, Gennaro
Pokorny, Jan
Pozderovic, Andrija
Räsänen, Janne
Robins, Elizabeth Naomi Clare
Roos, Yrjö Henrik

165

Salgò, András
Schwenke, Klaus Dieter
Sebecic, Blazenka
Sikora, Marek
Skrabka-Blotnicka, Teresa
Sousa, Isabel
Stetina, Jiri

Studer, Alfred
Subaric, Drago
Surowka, Krzysztof
Tiefenbacher, Karl
Tomasik, Piotr
Ugarcic-Hardi, Zaneta
Valentová, Helena

Viszkok, Ferenc
Vokk, Raivo
Wijngaards, Gerrit
Williams, Peter
Wilson, Philip

Sensor technology

Bosset, Jacques Olivier
Campanella, Luigi
Delgadillo, Ivonne
Di Natale, Corrado
Gramshaw, J.W.
Heiniö, Raija Liisa
Hitchcock, Christopher
Honikel, Karl Otto
Horne, David S.

Knowles, Michael Ernest
Mazzei, Franco
Pezacki, Wincenty
Ridgway, Christopher
Rymowicz, Waldemar
Smith, Linda Bernhardine Margaret
Smolinska, Teresa
Szoltysek, Katarzyna

Talou, Thierry
Trojan, Marek
Uccella, Nicola Antonio
Van Den Broek, Ad
Varadi, Maria
Von Wietersheim, Eugen
Wilson, Reginald

Sensory analysis

Applebye, Ulla
Bauer, Friedrich
Birch, Gordon G.
Boast, Martin
Botrè, Francesco
Brathen, Gudmund
Byrne, Briege Eileen
Calvo, Marta Maria
Clutton, David
Cotte, Virginie
Dawihl, Gerd
Dietrich, Helmut
Dirinck, Patrick
Drdak, Milan
Duran, Luis
Farmer, Linda
Feinberg, Max
Flowerden, Mary
Garncarek, Barbara
Gates, Leonard Michael
Gatti, Gian Carlo
Gertz, Christian
Gil, Ana Maria Pissarra C.
Gonzalez-Sanjose, Maria Luisa
Hakala, Mari
Häkkinen, Sari
Heiniö, Raija Liisa
Hills, Brian
Hruskar, Mirjana
Ilmoja, Kalle
Ingr, Ivo
Jarmoluk, Andrzej

Jirovetz, Leopold
Kedzior, Wladyslaw
Khokhar, Santosh
Klaarenbeek, Tineke
Kontominas, Michael
Kosicki, Zenon
Kostyra, Henryk
Kozlowska, Halina
Lambert, Michael
Lampolahti, Soili
Lapveteläinen, Anja
Leigh, Anthony
Leitner, Erich
Lesniak, Wladyslaw
Lingnert, Hans
Mikuschka, Gerhard
Molnar, Pal
Panovská, Zdenka
Parker, Jane
Peksa, Anna
Pfannhauser, Werner
Pisciotta, Gennaro
Pokorny, Jan
Pokorný, Jan
Pozderovic, Andrija
Pudil, Frantisek
Rogerson, Frank
Roozen, Jacques
Rops, Wichard
Saukko, Maire
Schlegel-Zawadzka, Malgorzata

Schmolck, Werner
Schreyen, Luc
Siegmund, Barbara
Skrabka-Blotnicka, Teresa
Smith, Linda Bernhardine Margaret
Sontag, Gerhard
Sousa, Isabel
Spooner, Martin John Richard
Stetina, Jiri
Suutarinen, Marjaana
Szoltysek, Katarzyna
Talou, Thierry
Tateo, Fernando
Timmermans, Eric
Toikkanen, Kari
Trojan, Marek
Troy, Declan J.
Uccella, Nicola Antonio
Ugarcic-Hardi, Zaneta
Uijttenboogaart, Theo
Vahcic, Nada
Valentová, Helena
Van Den Bosch, Gerrit
Velisek, Jan
Vella, Anthony
Vischer, Michaela
Vokk, Raivo
Von Wietersheim, Eugen
Williams, John Graham
Wucherpfennig, Karl

Spectroscopy

Adamantiadou, Sophia
Adams, J. Brian
Amaro Pinto, Rui Manuel
Amarowicz, Ryszard
Anklam, Elke
Aubourg, Santiago
Balling Engelsen, Soren
Beernaert, Hedwig
Beljaars, Paul
Belton, Peter
Benedito, Carmen
Berardo, Nicola
Berger, Ralf Günter
Betsche, Thomas
Birlouez, Inès
Brathen, Gudmund
Budde, Jürgen
Celik, Süeda
Clark, David
Copíková, Jana
Crossy, Paul
Curda, Ladislav
Damiani, Pietro
De Block, Jan
Delgadillo, Ivonne
Di Luccia, Aldo
Dietrich, Helmut
Donald, Athene
Dziuba, Jerzy
Ehlers, Dorothea
Empis, José
Evidente, Antonio
Faulds, Craig
Gaucheron, Frédéric
Gil, Ana Maria Pissarra C.
Gojmerac, Tihomira
Golja, Viviana
Goodall, David M.
Goodman, Bernard
Gramshaw, J.W.
Grenby, Trevor Hilary
Grenby, Trevor Hilary
Guillen, M. D.
Gunstone, Frank
Györi, Zoltán
Häkkinen, Sari
Hauser, Eugen J. B.
Henle, Thomas
Hey, Michael James
Hills, Brian
Hils, Arno K. A.
Iori, Renato
Ivanov, Kalintcho
Jirovetz, Leopold
Jones, James
Knieling, Ralph G.
Kontominas, Michael
Kovatcheva-Apostolova, Elena
Krska, Rudolf
Le Botlan, Denis
Luf, Wolfgang
Luneia, Roberto
Makris, Dimitris
Melchior Larsen, Lone
Meunier, Jean-Claude
Moreira Da Silva, Aida Maria
Munck, Lars
Obretonov, Tzvetan
Ooghe, Wilfried
Parr, Adrian James
Pisciotta, Gennaro
Rahali, Véronique
Ridgway, Christopher
Rizov, Nicolay
Roedig-Penman, Andrea
Rollin, Patrick
Russell, Wendy Roslyn
Rutledge, Douglas
Salgò, András
Schäfer, Karola
Scordaki, Alexandra
Sebecic, Blazenka
Sereno, Alberto M.C.
Sforza, Stefano
Steegmans, Monique
Steinhart, Hans
Suhaj, Milan
Suutarinen, Marjaana
Tauscher, Bernhard
Ternes, Waldemar
Thompson, Kenneth Clive
Tiefenbacher, Karl
Tomasik, Piotr
Tömösközi, Sándor
Van Renterghem, Roland
Varadi, Maria
Varey, Jane Elizabeth
Wang, Rui
Wijngaards, Gerrit
Wilson, Reginald
Winkeler, Heinz-Dieter
Yalcin, Erkan
Zyla, Krzysztof

Spices

Bachman, Stefania
Baines, David Allan
Beddows, Clifford G.
Beljaars, Paul
Berger, Ralf Günter
Boast, Martin
Brockmann, Rainer
Chiavaro, Emma
Dirinck, Patrick
Doncheva, Ivanka
Ehlers, Dorothea
Ekar, Igor
Farkas, Joszef
Farkas, Pavel
Fenwick, Gruffydd Roger
Frazier, Peter
Gates, Leonard Michael
Goodman, Bernard
Guillen, M. D.
Huopalahti, Rainer
Jarmoluk, Andrzej
Jirovetz, Leopold
Kerkvliet, Jacob
Khokhar, Santosh
Lambert, Michael
Liddle, Peter
Lugasi, Andrea
Marioleas, Panagiotis
Mosandl, Armin
Nehring, Ulrich P.
Ogaard Madsen, Jorgen
Pertoldi Marletta, Giuliana
Petz, Michael
Reglero, Guillermo J.
Sereno, Alberto M.C.
Spiro, Michael
Stevens, Roger
Surowka, Krzysztof
Tateo, Fernando
Ternes, Waldemar
Thornton, Raymond E.
Venskutonis, Petras Rimantas
Vieths, Stefan
Yanishlieva-Maslarova, Nedjalka
Zegota, Henryk

Spirits

Adamantiadou, Sophia
Beljaars, Paul
Campbell, Duncan J.
Carder, John
Clutton, David
Czarnecka, Maria
Czarnecki, Zbgniew
Dietrich, Helmut

Dirinck, Patrick
Hamilton, Colin A.
Hampton, Ian
Lehtonen, Pekka
Lesniak, Wladyslaw
Liddle, Peter
Martin, Gérard
Melzoch, Karel

Mosandl, Armin
Nowak, Jacek
Pfannhauser, Werner
Rogerson, Frank
Rychtera, Mojmír
Shackleton, Ronald
Stevens, Roger

Starch

Ahvenainen, Juha M.I.
Aura, Anna-Marija
Bachman, Stefania
Balling Engelsen, Soren
Beljaars, Paul
Berardo, Nicola
Betsche, Thomas
Bhat, Mahalingeshwara
Cooper, Julian
Czarnecka, Maria
Czarnecki, Zbgniew
Dirinck, Patrick
Donald, Athene
Eberle, Mike
Evidente, Antonio
Fisher, Leonard
Frazier, Peter

Garncarek, Barbara
Golachowski, Antoni
Grenby, Trevor H.
Hills, Brian
Home, Silja
Howard, Julie
Ilsbroux, Ingrid
Jones, Alan
Köksel, Hamit
Kosicki, Zenon
Kroh, Lothar W.
Le Botlan, Denis
Leszczynski, Waclaw
Lindhauer, Meinolf G.
Mac Dowall, James
Mischnick, Petra
Munck, Lars

Palasinski, Mieczyslaw
Räsänen, Janne
Rodziewicz, Anna
Roos, Yrjö Henrik
Sebecic, Blazenka
Seibel, Wilfried
Sikora, Marek
Slominska, Lucyna
Studer, Alfred
Tiefenbacher, Karl
Tomasik, Piotr
Van Eckert, Renate
Webb, Colin
Whitehouse, Brian
Williams, Peter
Wilson, Philip

Steroids

Amaro Pinto, Rui Manuel
Beernaert, Hedwig
Boskou, Dimitrius
Botrè, Francesco
Cerkvenik, Vesna
De Brabander, Hubert
Eisenbrand, Gerhard
Evangelisti, Filippo

Gojmerac, Tihomira
Jirovetz, Leopold
Lampi, Anna-Maija
Mc Donald, Mark
Piironen, Vieno
Rychtera, Mojmír
Saarinen, Niina
Schlett, Claus

Toivo, Jari
Tömösközi, Sándor
Van Peteghem, Carlos
Van Rhyn, Hans
Van Vyncht, Gery
Vanluchene, Eric
Zunin, Paola

Storage

Andrews, Anthony
Bauer, Friedrich
Bosset, Jacques Olivier
Calvo, Marta Maria
Collar Esteve, Concha
Czarnecka, Maria
Donald, Athene
Galic, Kata
Gaucheron, Frédéric
Golachowski, Antoni
Hajduk, Ewa

Häkkinen, Sari
Henle, Thomas
Henshall, David
Hood, Ted D.E.
Ilari, Jean-Luc
Jirovetz, Leopold
Klupsch, Robert N.
Krala, Lucjan
Lalljie, Sam
Lees, Ronald
Lingnert, Hans

Lisinska, Grazyna
Luf, Wolfgang
Molnar, Pal
Northolt, Martin
Palasinski, Mieczyslaw
Pastoriza, Laura
Peksa, Anna
Pezacki, Wincenty
Pipek, Petr
Przysiezna, Ewa
Räsänen, Janne

Ridgway, Christopher
Schmidt, Stefan
Sikorski, Zdzistaw

Smolinska, Teresa
Studer, Alfred
Surowka, Krzysztof

Talou, Thierry
Ugarcic-Hardi, Zaneta
Uijttenboogaart, Theo

Sugar, honey, sugar alcohols

Accorsi, Carla Alberta
Anklam, Elke
Beljaars, Paul
Bhat, Mahalingeshwara
Birch, Gordon G.
Bubnik, Zdenek
Christoph, Norbert
Cooper, Julian
Corradini, Claudio
Czarnecka, Maria
Czarnecki, Zbgniew
Eerikäinen, Tero
Ekar, Igor
Grenby, Trevor H.
Hampton, Ian

Hanewinkel-Meshkini, Susanne
Henle, Thomas
Hough, Leslie
Jörissen, Urban
Kerkvliet, Jacob
Klein, Erich
Lambert, Michael
Lees, Ronald
Mac Dowall, James
Martin, Gérard
Martinez-Castro, Isabel
Matissek, Reinhard
Moilanen, Raija
Molnar-Perl, Ibolya

Munck, Lars
Nöhle, Ulrich
Pietkiewicz, Jerzy Jan
Rychtera, Mojmír
Sikora, Marek
Skrökki, Leila Anneli
Speer, Karl
Steegmans, Monique
Studer, Alfred
Subaric, Drago
Van Rhyn, Hans
Vlater, Vladimír
Wang, Rui
Whitehouse, Brian

Sweeteners

Birch, Gordon G.
Budde, Jürgen
Chrysafidis, Dimitrios
Cooper, Julian
Corradini, Claudio
Dietrich, Helmut
Dostálová, Jana
Duran, Luis
Estelecki, Ilona
Garncarek, Barbara
Grenby, Trevor H.
Hanewinkel-Meshkini, Susanne

Hough, Leslie
Ilsbroux, Ingrid
Kovac, Spomenka
Kroyer, Gerhard Th.
Mac Dowall, James
Marioleas, Panagiotis
Matissek, Reinhard
Merlini, Lucio
Mikuschka, Gerhard
Panovská, Zdenka
Roozen, Jacques
Sakellariou, Christina
Salminen, Seppo Jaakko

Sass-Kiss, Agnes
Slominska, Lucyna
Spillane, William J.
Steegmans, Monique
Suhaj, Milan
Szoltysek, Katarzyna
Tateo, Fernando
Von Rymon Lipinski, Gert-Wolfhard
Whitehouse, Brian

Thermal processing

Aalbersberg, Willem Y.
Adams, J. Brian
Ashurst, Philip Roy
Aubourg, Santiago
Aura, Anna-Marija
Birlouez, Inès
Calvo, Marta Maria
Collar Esteve, Concha
Cotte, Virginie
De Block, Jan
Dostálová, Jana
Drdak, Milan
Frias, Juana
Gaucheron, Frédéric
Gertz, Christian
Häkkinen, Sari

Hegedusic, Vesna
Henle, Thomas
Henshall, David
Hills, Brian
Iciek, Jan
Jägerstad, Margaretha
Khokhar, Santosh
Kmiecik, Waldemar Andrzej
Kontominas, Michael
Kozlowska, Halina
Kroh, Lothar W.
Luf, Wolfgang
Matuszek, Tadeusz
Mills, Elizabeth Naomi Clare
Mlotkiewicz, Jerzy
Northolt, Martin

Novakovic, Predrag
Obretonov, Tzvetan
Olano, Agustin
Palka, Krystyna
Pezacki, Wincenty
Pipek, Petr
Pischetsrieder, Monika
Pokorný, Jan
Rutledge, Douglas
Schmitt, Rudolf
Schreyen, Luc
Steinbuch, Erwin
Studer, Alfred
Ternes, Waldemar
Tiefenbacher, Karl
Tykkyläinen, Paavo

Van Boekel, Tiny
Van Den Bosch, Gerrit

Williams, John Graham
Wucherpfennig, Karl

TLC

Adamantiadou, Sophia
Amarowicz, Ryszard
Anklam, Elke
Baltes, Werner
Barron, Luis Javier Rodríguez
Beernaert, Hedwig
Beljaars, Paul
Berger, Ralf Günter
Betsche, Thomas
Calvo, Marta Maria
Corradini, Claudio
Damiani, Pietro
De Brabander, Hubert
Demopolous, Constantinos A.
Di Luccia, Aldo
Doncheva, Ivanka
Eisenbrand, Gerhard
Evidente, Antonio

Ferrara, Lydia
Filip, Vladimir
Gojmerac, Tihomira
Graille, Jean
Haffke, Helma
Henle, Thomas
Ilmoja, Kalle
Jirovetz, Leopold
Knieling, Ralph G.
Kombal, Ralph
Kovac, Spomenka
Krkoskova, Bernadetta
Kroh, Lothar W.
Laakso, Päivi
Luneia, Roberto
Makris, Dimitris
Neicheva, Anastasia
Rizov, Nicolay

Roedig-Penman, Andrea
Rollin, Patrick
Schäfer, Karola
Schmidt, Stefan
Scordaki, Alexandra
Simon-Sarkadi, Livia
Sinigoj-Gacnik, Ksenija
Stanoeva, Elena
Steiner, Ingrid
Stempiewicz, Regina
Studer, Alfred
Ternes, Waldemar
Toivo, Jari
Trojan, Marek
Van Vyncht, Gery
Yanishlieva-Maslarova, Nedjalka
Ziino, Marisa

Toxic trace elements

Baars, Aalbert Jan
Bauer, Ulrich
Beernaert, Hedwig
Betsche, Thomas
Branch, Simon
Burini, Giovanni
Bykowski, Piotr Jan
Corvi, Claude Albert
Deelstra, Hendrik
Dettweiler, Gerd
Doganoc, Darinka Zdenka
Ducauze, Christian J.
Ellen, Geert
Enders, Peter W.
Györi, Zoltán
Haffke, Helma
Henshall, David
Honikel, Karl Otto
Howick, Chris

Ilmoja, Kalle
Kalac, Pavel
Kellner, Vladimir
Klein, Bernard
Kovac, Spomenka
Lach, Günter
Lalljie, Sam
Mandic, Milena L.
Pavelka, Jiri
Pertoldi Marletta, Giuliana
Pezacki, Wincenty
Pfannhauser, Werner
Pollmer, Udo
Ristow, Reinhard
Rizov, Nicolay
Sapunar-Postruznik, Jasenka
Schlatter, Christian
Schlegel-Zawadzka, Malgorzata

Schlett, Claus
Schneider, Rüdiger
Sebecic, Blazenka
Sinigoj-Gacnik, Ksenija
Skibniewska, Krystyna A.
Stelz, Alice
Thompson, Kenneth Clive
Thornton, Raymond E.
Tiusanen, Sirkka
Van Den Bosch, Gerrit
Van Der Schee, Henk A.
Van Vyncht, Gery
Varey, Jane Elizabeth
Von Wietersheim, Eugen
Wiggins, Edgar Hugh
Wucherpfennig, Karl
Zidaric, Metka
Zoller, Otmar

Toxicology

Amaro Pinto, Rui Manuel
Baars, Aalbert Jan
Battaglia, Reto
Bauer, Ulrich
Bazulic, Davorin
Beljaars, Paul
Botrè, Claudio
Botrè, Francesco

Brunn, Hubertus
Burini, Giovanni
Corvi, Claude Albert
Daphi-Weber, Juliane
Dettweiler, Gerd
Dos Santos Baptista, Bráulio
Eisenbrand, Gerhard
Ellen, Geert

Gojmerac, Tihomira
Hahn, Harald
Hermus, Rudolph J.J.
Herraiz Tomico, Tomas
Holst, Birgit
Howick, Chris
Hrncirik, Karel
Ilmoja, Kalle

Jirovetz, Leopold
Kerkvliet, Jacob
Knieling, Ralph G.
Knowles, Michael Ernest
Lalljie, Sam
Landsiedel, Robert
Lautenbacher, Lutz-Michael
Mazzei, Franco
Noteborn, Hubert
O' Neill, Ian
O´Brien, John
Pischetsrieder, Monika
Pollmer, Udo
Restani, Patrizia
Rizov, Nicolay
Salminen, Seppo
Sapunar-Postruznik, Jasenka
Schlegel-Zawadzka, Malgorzata
Schmidt, Heinz
Schrenk, Dieter
Skibniewska, Krystyna A.
Stelz, Alice
Suhaj, Milan
Szilágyi, Szilárd
Ternes, Waldemar
Thornton, Raymond E.
Van Peteghem, Carlos
Velisek, Jan
Zoller, Otmar

Trace elements

Amaro Pinto, Rui Manuel
Baars, Aalbert Jan
Bauer, Ulrich
Beddows, Clifford G.
Beernaert, Hedwig
Beljaars, Paul
Betsche, Thomas
Bontempelli, Gino
Branch, Simon
Burini, Giovanni
Bykowski, Piotr Jan
Corvi, Claude Albert
Crossy, Paul
De Brabander, Hubert
Deelstra, Hendrik
Dettweiler, Gerd
Doganoc, Darinka Zdenka
Ducauze, Christian J.
Ellen, Geert
Enders, Peter W.
Favretto, Luciano
Gabrielli Favretto, Luciana
Gaucheron, Frédéric
Goodman, Bernard
Györi, Zoltán
Haffke, Helma
Henshall, David
Honikel, Karl Otto
Horvatic, Marija
Howick, Chris
Ilmoja, Kalle
Juarez, Manuela
Kalac, Pavel
Kellner, Vladimir
Khokhar, Santosh
Kieffer, Felix
Klapec, Tomislav
Klein, Bernard
Knieling, Ralph G.
Kovac, Spomenka
Kucera, Jiri
Lach, Günter
Lalljie, Sam
Lautenbacher, Lutz-Michael
Leibetseder, Josef
Long, Alan
Mandic, Milena L.
Pavelka, Jiri
Pertoldi Marletta, Giuliana
Pezacki, Wincenty
Pfannhauser, Werner
Pollmer, Udo
Primorac, Ljiljana
Ristow, Reinhard
Rizov, Nicolay
Rymowicz, Waldemar
Sandberg, Ann-Sofie
Sapunar-Postruznik, Jasenka
Schlatter, Christian
Schlegel-Zawadzka, Malgorzata
Schlett, Claus
Schneider, Rüdiger
Sebecic, Blazenka
Sinigoj-Gacnik, Ksenija
Skibniewska, Krystyna A.
Stelz, Alice
Thompson, Kenneth Clive
Thornton, Raymond E.
Tiusanen, Sirkka
Tomasik, Piotr
Van Den Bosch, Gerrit
Van Der Schee, Henk A.
Van Dokkum, Wim
Van Vyncht, Gery
Varey, Jane Elizabeth
Von Wietersheim, Eugen
Wiedner, Peter
Wiggins, Edgar Hugh
Wucherpfennig, Karl
Zidaric, Metka
Zoller, Otmar

Transgenic foods

Anklam, Elke
Betsche, Thomas
Decaris, Bernard
Drdak, Milan
Gatti, Gian Carlo
Gunstone, Frank
Henle, Thomas
Herman, Lieve
Ilsbroux, Ingrid
Jörissen, Urban
Kedzior, Wladyslaw
Kostyra, Henryk
Kucera, Jiri
Marchelli, Rosangela
Martelli, Aldo
Melzoch, Karel
Noteborn, Hubert
Nowak, Jacek
Otteneder, Herbert
Parr, Adrian James
Pezacki, Wincenty
Siezen, Roland J.
Thornton, Raymond E.
Waldron, Keith
Zyla, Krzysztof

Vitamins

Ager, Elaine
Anklam, Elke
Aust, Olivier
Beddows, Clifford G.
Beernaert, Hedwig
Beljaars, Paul
Betsche, Thomas
Birlouez, Inès
Böhm, Volker
Brathen, Gudmund
Burini, Giovanni
Chrysafidis, Dimitrios
Davidek, Jiri
Frias, Juana
Fryer, John
Galluser, Anita
Gojmerac, Tihomira
Graille, Jean
Hägg, Margareta
Havenaar, Robert
Heinonen, Marina
Hermus, Rudolph J.J.
Holasova, Marie

Holmer, Gunhild
Honkavaara, Markku
Hrncirik, Karel
Jägerstad, Margaretha
Khokhar, Santosh
Knieling, Ralph G.
Kovatcheva-Apostolova, Elena
Lampi, Anna-Maija
Lautenbacher, Lutz-Michael
Lisiewska, Zofia Barbara
Long, Alan
Lossonczy Von Losoncz, Thomas
Mandic, Milena L.
Masková, Eva
Moilanen, Raija
Mörsel, Jörg-Thomas
Moser, Ulrich
Murkovic, Michael
Nortvedt, Ragnar
Ollilainen, Velimatti
Piironen, Vieno
Pollmer, Udo

Popken, Anne M.
Ragotzky, Klaus
Ribarova, Fanny
Rizov, Nicolay
Roedig-Penman, Andrea
Schwack, Wolfgang
Sebecic, Blazenka
Sheppard, Peter D.
Szilágyi, Szilárd
Tauscher, Bernhard
Thornton, Raymond E.
Trojanowska, Krystyna
Vahteristo, Liisa
Van Den Bosch, Gerrit
Van Vyncht, Gery
Vanluchene, Eric
Velisek, Jan
Vidal-Valverde, Concepcion
Waldron, Keith
Wilska-Jeszka, Jadwiga
Winkler, Johanna
Zegota, Henryk

Water

Adamantiadou, Sophia
Amaro Pinto, Rui Manuel
Balling Engelsen, Soren
Bauer, Ulrich
Belton, Peter
Birch, Gordon G.
Bontempelli, Gino
Brathen, Gudmund
Campanella, Luigi
Christoph, Norbert
Crossy, Paul
Dechow, Arndt
Ducauze, Christian J.
Erning, Dieter
Gojmerac, Tihomira
Hägg, Margareta
Hauser, Eugen J. B.
Hey, Michael James

Holandová, Katerina
Hood, Ted D.E.
Ilmoja, Kalle
Kontominas, Michael
Krska, Rudolf
Le Botlan, Denis
Luf, Wolfgang
Mac Dowall, James
Miskiewicz, Tadeusz
Moreira Da Silva, Aida Maria
Moret, Ivo
Northolt, Martin
Otteneder, Herbert
Pezacki, Wincenty
Pisciotta, Gennaro
Räsänen, Janne
Richter, Timo
Saltmarsh, Mike

Schlett, Claus
Schmid, Erich R.
Shackleton, Ronald
Siebers, Johannes
Skrökki, Leila Anneli
Spiro, Michael
Steiner, Ingrid
Stelz, Alice
Studer, Alfred
Tauscher, Bernhard
Thompson, Kenneth Clive
Tiusanen, Sirkka
Topalova, Ivanka Chrstova
Van Vyncht, Gery
Wiedner, Peter
Wiggins, Edgar Hugh
Wilson, Philip
Winkeler, Heinz-Dieter

Wine

Aura, Anna-Marija
Beddows, Clifford G.
Beljaars, Paul
Bontempelli, Gino
Botrè, Francesco
Buglass, Alan J.
Carder, John
Christoph, Norbert

Clutton, David
Coimbra, Manuel António
Delgadillo, Ivonne
Demopolous, Constantinos A.
Di Natale, Corrado
Dietrich, Helmut
Dirinck, Patrick
Dos Santos Baptista, Bráulio

Drdak, Milan
Ducauze, Christian J.
Enders, Peter W.
Farkas, Pavel
Gonzalez-Sanjose, Maria Luisa
Häkkinen, Sari
Hamilton, Colin A.
Hils, Arno K. A.

Jirovetz, Leopold
Kaufmann, Anton
Klupsch, Robert N.
Lehtonen, Pekka
Liddle, Peter
Lugasi, Andrea
Mac Dowall, James
Makris, Dimitris
Martin, Gérard
Mazzei, Franco
Melzoch, Karel

Moret, Ivo
Ooghe, Wilfried
Otteneder, Herbert
Pisciotta, Gennaro
Popov, Dimitre
Pozderovic, Andrija
Ristow, Reinhard
Rogerson, Frank
Rops, Wichard
Rychtera, Mojmír
Sass-Kiss, Agnes

Shackleton, Ronald
Simon-Sarkadi, Livia
Thornton, Raymond E.
Topalova, Ivanka Chrstova
Valentová, Helena
Vischer, Michaela
Williams, Mervyn
Winterhalter, Peter
Wucherpfennig, Karl

Printing (Computer to Film): Saladruck, Berlin
Binding: Stürtz AG, Würzburg